MANAGEMENT
OF
INTERNATIONAL
NETWORKS

Cost-Effective Strategies for the
New Telecom Regulations and Services

3 March 2004

To my voice-friend Chris,

Some reading on regulation of
your favourite new services, but
even more to build your own
international network of people
that are happy fun working together

Florio van der Broek

MANAGEMENT
OF
INTERNATIONAL
NETWORKS

*Cost-Effective Strategies for the
New Telecom Regulations and Services*

Floris van den Broek

CRC Press
Boca Raton   London   New York   Washington, D.C.

Acquiring Editor: Dawn Mesa
Project Editor: Sylvia Wood
Cover design: Dawn Boyd

Library of Congress Cataloging-in-Publication Data

Broek, Floris van den
    Management of international networks: cost-effective strategies
for the new telecom regulations and services / Floris van den Broek.
        p. cm.
    Includes bibliographical references and index.
    ISBN 0-8493-0739-2 (alk. paper)
    1. Telecommunication--management. I. Title.
HE7661.B76 1999
    658'.05467—dc21
    DNLM/DLC
    for Library of Congress
                                                                99-32369
                                                                     CIP

Preface

In recent years, a giant telecommunications boom has emerged. The rise of the Internet, the liberalization of telecommunications regulation, and the appearance of many new telecommunications companies are just a few of signs of the fast-growing need for communication among people around the globe.

As the science of network management is relatively new, very little research has been done in that area, and most of the decisions on initial investments to set up a new system are still made without knowledge of the management aspects. Internationalization of networks adds an extra factor of complexity that can have an enormous impact on their costs and effectiveness. A lack of knowledge of management of international networks often leads to high management costs for organizations. This, combined with the fast-rising demand for international communication by users of the networks, can result in costs getting completely out of control. This book explores international aspects that influence management of networks.

The research in this book was conducted as part of the research program of Information Strategy and Management of Information Systems of the Delft University of Technologyand supported by the Columbia Institute of Tele Information at Columbia University in New York. The program in Delft has included several doctoral projects such as Modeling Change Management of Evolving Heterogeneous Networks, Evaluation of Information Systems Proposals, Management of Distributed Data in Distributed Environments, and Information Systems Management in Complex Organizations.

The research proposes a model, the cost-effective management model, which helps us understand the relationship among regulatory environment, telecommunications services offering and cost-effective management. The research was based on a survey, an explorative case and two test cases, as well as practical experience in large telecommunications companies in Europe and in the United States.

Acknowledgments

The process of research and writing of the book became international in itself, as part of the research was conducted in the Netherlands and part in the United States, following my move to the U.S. in 1995. Thanks to the flexibility of my promoter, it was possible to continue the research based in the U.S.

An "international network" of people in both the Netherlands and the U.S. have helped me with the research for this book. First, for his excellent support, I would like to thank Maarten Looijen, who somehow managed to bridge the gap between the Netherlands and the U.S. to accommodate my travel schedules. Eli Noam of Columbia University in New York gave me close guidance and was invaluable from the telecommunications regulations point of view as a second doctoral advisor. In the same area, Willem Korthals Altes and Mike Botein of the New York Law School made essential contributions. Strong support also came from Jens Arnbak, president of OPTA, the regulatory body of the Netherlands; Erkki Likanen, the European Commissioner for telecommunications; and Theo Bemelmans of Eindhoven University of Technology. Kornel Terplan, Wim Verdonck and Bruce Egan lent me their unique experiences from their management consulting practice. I thank Louis van Hemmen, Rob Mersel and Cees de Wijs, doctoral-degree students at the Delft University of Technology, for intense and exciting academic discussions, and Esther Rombaut of Columbia University for her key role in the test cases with Lawrence Yeung of the United Nations and John Stabrylla of Calvin Klein Cosmetics. The telecommunications-industry input came from Jan Sander (KPN Telecom), Dick Witstok, Mike Stoll (Lucent Technologies), Bob Baulch, Cathy Martine (AT&T), Marc Hendrickx (INS), and Jo van Gorp (Level 3). Finally, my wife Eva provided continuous support without which I could not have endured the endless evenings and weekends on this project.

Floris van den Broek
Amsterdam, July 1999

Introduction

Management of a network can be a costly activity. When a network is international (i.e., it crosses international borders) this is even more applicable. This book helps to guide strategic decisions that management of an international network requires—strategic decisions that include in what country to expand a network or place a hub for international traffic, but that can be as far-reaching as where to place a regional headquarters. As an illustrative example: Hewlett-Packard moved its European Headquarters from Brussels to London in the early '90s because of the difficulties with regulations and with obtaining high-quality, low-cost communications services for its network.

What's Special about International Networks

By means of analysis of trends, surveys, and explorative cases, several points are identified that need particular attention when managing an international network. Examples are:

- The different dominant telecommunications operators in a country
- Regulation for telecommunications varies in different countries
- The different languages often spoken in other countries
- The varied offerings of telecommunications services in different countries

Regulatory Environment and Telecommunications Services Offering in Countries

Two of the aforementioned attention points are particularly important for management of international networks:

1. The regulatory environment for telecommunications services in a country
2. The telecommunications services offered in a country

These attention points are each described in detail and expressed in models with which they can be quantified. The result is a quantitative assessment of the regulatory environment and of the telecommunications services offered in a country.

Measuring Cost-Effective Management

To discover how these attention points exactly influence management of a network, we use a measurement tool for assessing what we will call "cost-effective management" of a network. Employing both reference materials and practical experience, an easy-to-use measurement is developed that both more clearly defines what is meant by cost-effective management and how it can be quantified.

Applying the Models

In a few practical cases of international networks, the models are applied and the influence of both the regulatory environment and telecommunications-services offerings on the cost-effective management of a network is assessed.

The cases are carried out in Lucent Technologies, United Nations Development Program and Calvin Klein Cosmetics Company. Certain parts of the regulatory environment model offer more predictive value than others. For example, in a given country, the independence of the regulatory body and the licensing process (that the telecommunications license be awarded in an open, fast, and nondiscriminatory way) seem to have a more than usual correlation with the ability to cost-effectively manage a network in that country. Also, the presence of "universal service" (the obligation to provide service to all users of a network on the same terms and conditions), proves to be an indicator of a good country to establish (or place a hub for) a network.

Conclusions

Assessing the regulatory environment and the telecommunications services offering of a country with the models described can forecast the possibilities to cost-effectively manage a network in that country with reasonable confidence. Such an assessment can offer invaluable help with strategic questions, such as where to place hubs for traffic in an international network, in what country to expand the network, what services to obtain from which operator in order to run the network most cost-effectively, and many others.

Author

Floris van den Broek graduated from the University of California, Berkeley with a Master's in business administration (1990). He holds a *Doctoraal* (MSc) in computer science (1988) from University of Nijmegen (graduation at Eindhoven University of Technology) in the Netherlands as well as *Propedeuse* degrees in law and economics from the universities of Nijmegen and Amsterdam, respectively. He earned his Ph.D. with research on Management of International Networks at Delft University of Technology in 1999.

He is currently managing director, the Netherlands, for Level (3) Communications. From 1990 to 1998, at Lucent Technologies and AT&T, he held various positions such as international sales manager, human resources manager Eastern Europe, and European product manager for switching systems in the regional headquarters in the Netherlands. He worked in the U.S. for three and a half years as international sales director global accounts at Lucent Technologies, and as multilateral services director for AT&T, responsible for international networks for nontraditional services. He is also a board member for several Internet startups.

Dr. Van den Broek is author of a number of articles on management of international networks, of which two appeared in the *International Journal of Network Management* in spring and summer 1997. He is also associate editor of the *Telecommunications Handbook*, published in 1999 by CRC Press.

Table of Contents

1 Management of International Networks

More and more, international communications networks are becoming part of our everyday lives. Whether we reserve an airplane seat through a central reservation agency or get money from a banking machine in the country in which we are vacationing, we are dependent on international networks. Airlines, banks, hotel chains, and other international organizations consider networks to be the most essential part of their business. The rise of the Internet is an example of how individuals are reaping the benefits of direct access to an international network. Many people now find it normal to quickly access information from the other side of the world by means of a computer, an operation that would have been considered magic just a few years ago.

International networks can consist of computers in different countries that are connected with each other, but can also be effected via a telephone network, where a person in one country can call someone in another country and be connected by just pressing a few buttons on a telephone set.

1.1 INTRODUCTION

International networks need to be designed, built, and managed. Studies by the Index Group [Treacy 1989, Gartner 1994] show that the management of networks is the most expensive* part of these activities. One of the reasons for this is the cost of international telecommunications services, on which organizations are spending large and fast-increasing amounts of money.

The demand for international communications is growing rapidly with the introduction of new services and as more and more people begin to make use of the services of international networks. On the other hand, the cost per unit of information (whether expressed in minutes or bits or other units), is quickly shrinking because of technological advancements such

---

* Research of an index group [Treacy 1989] shows that over a five-year period, the costs of management of a network account for two thirds of the total costs of a network. Only one third of the total costs of the network is spent for the purchase of the network equipment and software.

as the increasing capacity of fiber-optic cables [Frieden 1995]. As yet, the growth in demand for telecommunications services outpaces the price decrease, and therefore the average telecommunications bill for international organizations continues to rise rapidly. In spite of their efforts to manage the cost of international communications, international organizations often find that the environment in many countries in which they operate strongly limits their ability to do so. Another reason for the high cost may be the lack of availability of the right services in the sections of the countries where the network operates.

In countries with competition in telecommunications, there are market forces that can influence the prices, quality, and the telecommunications-services offering [Lamberton 1997], but this is not always the case. In addition to market forces, government measures can determine what services are offered and at what price. Price and quality of telecommunications services are often factors that influence the possibilities of international organizations to conduct business or government activities, or maintain an international network in that country.

Right now, the telecommunications environment in many countries is changing or about to change drastically. Governments are liberalizing the offering of services for international networks, privatizing formerly government-owned telecommunications operators and introducing competition. In the European Union, for example, 1998 was an important year, as it marked the opening of the telecommunications markets to competition in several of the member states, with the other member states slated to follow before the end of 2002. On a worldwide scale, in 1997 the GATS (General Agreement on Trade and Services) agreement was signed, in which 69 member states of the World Trade Organization (WTO) ([Drake 1997][Sisson 1997][Oliver 1998]), agreed to a time schedule to open their countries to competition in the market for telecommunications services. This means that the signatory countries of the GATS agreement will each open their markets at a predetermined time between the years 1998 and 2002.

The strong increase in demand for international network services on one hand and the rapid changes in the telecommunications environment in many countries on the other, call for a closer examination of international networks. This chapter will focus on the reasons, means, and approach (why, what, how?) in such an examination and will also give a brief overview of current literature on the management of networks.

1.2   INTERNATIONAL NETWORKS

An international network crosses international borders. International networks are often characterized as no different from a national network—particularly tempting in a world where country boundaries are disappearing every day. Indeed, some countries cooperate in such a way that the environments for the international networks appear quite similar to their neighbor countries (some countries in the European Union, for example, are attempting to create similar telecommunications laws), but in reality, differences, such as variances in the regulatory environment, language, or dominant telecommunications-services suppliers still exist and are sometimes becoming more pronounced because of the varied pace of regulatory change in various countries.

We should define what we mean by a network and an international network. There are several definitions for the word network in the area of electronic communications. For our purposes, a network is *a composition of communication devices and links that connect at least two nodes that consist of hardware and software. These connected communications devices and links perform interactions between nodes using exact prescriptions, including protocols* [VanHemmen 1997].

In this volume, an international network is defined as *a network that operates in at least two different countries, with a management organization (coordinating body) in each of the countries responsible for the management of that part of the network that lies in that country.*

A topology of a network shows the locations of nodes and links of a network. Figure 1.1 shows two examples of international networks with different topologies. One of the differences between these two examples is the number of links crossing an international border (international links). The left network has four international links, whereas the right network has only two. The total number of links in these examples is equal. Although designing a network topology is not the subject of this book, knowledge of cost effectiveness of managing international links vs. noninternational links can influence the choice for a topology. Apart from different cost effectiveness, the topologies may also have different performance and reliability characteristics and redundancies, should problems occur such as a broken link caused by a ruptured cable.

The examples in Figure 1.1 show only links that connect nodes in neighboring countries. However, links in international networks may also connect nodes in countries that are not neighbors and the links may therefore skip a

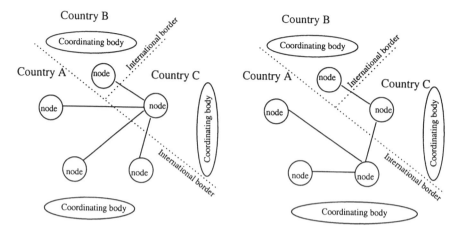

FIGURE 1.1 Examples of international networks with different topologies.

series of countries before ending at a node. The countries that are skipped (and therefore have no node) are, in this book, not taken into consideration.*

Management

Management is defined to include the activities of "control" and "maintenance" [Looijen 1998]. Management is therefore also referred to as "management, control, and maintenance (MCM). Here we have chosen to use the word management to represent all three of the activities and the "control" and "maintenance" parts should be considered included.

The definition of an international network suggests that there are different organizations (coordinating bodies) responsible for management in each of the countries. From an organizational point of view, this can mean that the coordinating bodies are logically in the same organization in the organization diagram. There are, however, always some management activities that are carried out separately on each side of an international border and that are therefore viewed as carried out by several "coordinating bodies." For example, a group of specialists in an Information Technology (IT) department might support multiple countries, but would maintain on-site assistants in each of the countries.

---

* Not considering the skipped countries is generally accepted to be the most accurate method. Physical links that carry the information between the nodes can be, e.g., satellite (which literally skips countries that lie in between) or fiber cable, which may skip countries, depending if it is laid under the sea or over land.

Trends in International Networks

Four trends affecting international networks have been found in practice and in literature [Terplan 1992, Frieden 1995, Ananthanpillai 1997].

1. Increased scale and globalization
2. Changing regulatory environments
3. A widening range of telecommunications services being offered
4. The trend to subcontract management outside the organization (outsourcing)

Increased Scale and Globalization of Networks

International networks are in rapid development and a concurrent demand for international communications has risen sharply. Consumers in various parts of the world are asking for more and more services across international borders. The number of countries that are connected to international networks has increased drastically. When an organization starts to build its international network in just in a few countries, the people in parts of the organization in the rest of the world often realize that they would also like access to the same information. A good example is the rise of corporate intranets, which transport more and more internal reference information and messages (such as electronic mail) among offices of the same company. Working together effectively for different parts of the organization in different countries will soon require access to an intranet (which, if linking offices in different countries, is an international network) in all local offices of the organization.

A Changing Regulatory Environment

Liberalization of telecommunications markets around the world currently is an important trend that helps change international networks. Liberalization usually means that telecommunications services may be offered by multiple parties and not, as was traditionally the case, only by one monopolistic telecommunications company or ministry (public telecommunications operator, or PTO). Also, PTOs are increasingly privatized, which makes them independent of governments and gives them more freedom to determine their own market strategy [Noam 1994].

A Widening Range of Telecommunications Services

Until recently, the choice of telecommunications services in many countries, in particular the services that were offered internationally (cross-border), was very limited. Public switched telephone network, also called "voice telephony," leased lines and some limited data services, such as low-speed packet-switched data services, often composed the total offering. These international services were (and still are) much more expensive than similar services being offered within a country [Frieden 1995, Zarley 1997]. Currently, the array of available international services is expanding rapidly in most countries, including international ISDN (integrated services digital network), often used for video conferencing and fast, low-cost data transfer, frame relay; ATM (asynchronous transfer mode); and TCP/IP (transaction control protocol/Internet protocol) connections for Internet traffic. Furthermore, a series of cross-border wireless services, such as GSM (global system for mobile telecommunications) with roaming has emerged.

The prices for these services have also been lowered considerably and fixed-rate pricing of the Internet connections have helped to make international services through the Internet extremely popular. In the future, an even wider range of telecommunications services will be offered, as their distribution will become easier because of new technology, such as coaxial cable, compression technologies, and wireless technologies, to deliver the services to the end-user.

Outsourcing
Outsourcing is playing an increasingly important role in the IT industry, with an increase in popularity for management of networks [Zarley 1997]. While the international network has traditionally been so essential to most of the larger international organizations that they have so far been very careful to maintain control over their own networks, outsourcing companies are now actively assuming several of the management segments of organizations' international networks. Some of the trends (widening range of telecommunications services being offered and a changing regulatory environment) that result in more suppliers and services to choose from, have led to a need for specialized knowledge of the purchase of telecommunications services [Roussel 1996].

These trends are not alone. Others that can be found in literature are, e.g., standardization of communications protocols and convergence of networks into one technology. Standardization is often stimulated by governments and industry, but is sometimes also avoided as governments try to discourage a concentration of power.

1.3 REASONS FOR THE RESEARCH (WHY?)

Managing an international network is different from managing a national network. Why should it be important to learn more about the management of international networks? Since an international network crosses borders and operates in various countries, particularities of the countries, such as the language, telecommunications-services suppliers and the telecommunications regulations can influence the management of an international network. Understanding the particularities of and impact on the management of the network helps to organize such management for maximum cost effectiveness. Owning and managing an international network involves a considerable investment in resources such as people, equipment, and external facilities, which could be used in a more efficient and effective way with knowledge of exactly what influences cost-effective management. Many organizations that own and manage international networks can benefit from knowing what factors influence the cost-effective management of the network in order to improve performance and lower costs of their networks.

Developments in society as well as in international organizations also show a need for more understanding of what exactly influences management of an international network, as is shown in the following.

The Role of Governments and Regulatory Organizations

Governments and regulatory organizations have as some of their main tasks to make and enforce regulations to improve quality of life and stimulate economic growth in their country or territory. By understanding how the regulatory environment would influence cost-effective management of a network, regulations can be made and the set of telecommunications services being offered can be adjusted. Doing so may benefit the users of telecommunications services in that country.

Governments and regulatory organizations can be supported by knowledge about the influence of their regulations on cost-effective management of networks in their country. This enables them to select regulations that have a positive influence on management of international networks and thus on the attractiveness for international organizations to establish a (regional) telecommunications headquarters, also called a "hub." A hub is a node with a position in the network that acts as a concentration point. Arnbak [Arnbak 1999] stresses the importance of understanding the technical dynamics and characteristics for every governmental body that makes policies and rulings.

Reasons for International Organizations

International organizations in telecommunictions, such as AT&T, MCI/World-Com, Qwest and Level(3) Communications, may manage both international networks for their own use, as well as international networks of other organizations. Other organizations such as multinationals manage networks for themselves. Political developments in most of the world are aimed toward ending many of the monopolies of telecommunications operators over the provision of telecommunication services. The environment of a manager of an international network becomes much more complex. A timely insight into the influences that affect multinational organizations managing their international networks and purchasing telecommunications services is essential for international organizations. The large international organizations together spend more than $100 billion per year—a figure that is growing at 15% annually [AT&T 1997]—on telecommunications services for their networks that they buy from other parties, excluding equipment and in-house cost for management control and maintenance.

International organizations must answer important questions and address the need to make strategic decisions that could influence the cost-effective management of their networks. Some *strategic questions* might be:

- Assuming the country choice is flexible, how does an organization choose a country for establishing a new part of its international network?
- How does the organization design an international network topology, including border-crossing links, so that the international network can be managed cost effectively?
- How is a country chosen as a location for a "hub" to concentrate international traffic among countries in the most cost-effective way?

Example of the Influence of International Aspects on Management of Networks

Real-life examples are illustrated by examining experiences with communications managers of international companies as described first in a consultant report [McKinsey 1993]. In 1992, Wal-Mart, a large U.S. retailer of household goods wanted to expand its business into Europe. Wal-Mart, which uses a completely automated logistics system as its main differentiator in the market, has super-efficient logistics that result in few stockouts, meaning few disappointed customers and low inventory costs. One of the managers of its international network indicated that management of Wal-Mart's network can be managed with lower costs in the U.S. than in Europe. One reason is that international leased-line prices are much higher in Europe (10–15 times) even in the year 2000, which made the company's business plan look so negative that the company was forced to decide not to continue in the European market. The other main cost factor in the management of the network was personnel, since costs for personnel turn out to be higher in most European countries than in the U.S. However, the difference was far less than the difference between the leased-line prices. Wal-Mart's main advantage could not be achieved cost effectively in Europe. Similar experiences and business applications can be found in other literature [Bartholomew 1997].*

Another example can be found in the experience of Hewlett-Packard, which moved its European headquarters from Brussels to London in the early '90s because of the difficulties in obtaining high-quality, low-cost telecommunications services for the company's network.

1.4 A MODEL FOR MANAGEMENT OF INTERNATIONAL NETWORKS

The factors taken into account in choosing a model are:

- Existing literature on international aspects of a country for management of an international network
- The four trends mentioned in section 1.2
- The three strategic questions discussed in section 1.3

Existing literature on international aspects as basis for the model

There is very little literature that addresses management of an international network as "attention points" [Liebmann 1995]. The attention points listed by Liebmann for managing an international network are described as follows:

- Geographical size of the network
- Crossing of multiple time zones by the network
- Different languages spoken in the countries where the network is operated
- Different social cultures of the countries where the network is operated

---

* Possibly the entry into Europe would have been eased by careful planning of the network, such as avoiding too many lines in the "higher-price leased line countries" or too many lines crossing borders, using modern equipment that compresses information and saves capacity, but even those costs would have been too high.

- Offering of international telecommunications services
- "Product politics" influencing the choice of hardware and software as suppliers (choosing telecommunications equipment is in part determined by suppliers that do not service/sell all equipment in every country)

Closer examination reveals that the first four of Liebmann's attention points are not mentioned solely because the network is international, but take into account other reasons that can also exist within a country. For example, the first attention point, the geographical size of the network and the second point, the crossing of multiple time zones by the network, are not particular to an international network. Networks that exist in the U.S. are not international, yet cover much more territory than many networks in Europe that are international. Very often, moving to a different country means that a different language is spoken as stated in attention point 3, but the country and the language are not necessarily coupled. Also, cultures, as stated in attention point 4, may change when crossing country borders, but they may also be very different within a country.

Attention point 5 in Liebmann's list, offering of international telecommunications services, has support in literature* as it also includes price [Frieden 1995, OECD 1997, Van den Broek 1997].

Attention point 5, "product politics" is particularly important when the organization responsible for management has standardized on certain types of equipment or if the presence of particular equipment is necessary at the foreign nodes to perform services correctly. Literature [Zarley 1997] shows a strong increase in demand for delivery of the same equipment across borders. Network managers ask not only that their standard equipment be delivered in all international locations, but also that similar services from the equipment suppliers, such as help-desk, warranty support, and system integration, be available. Suppliers to organizations with international networks establish equipment service offices, or cooperate with other equipment service vendors to be able to provide global support.

Recently, equipment suppliers have become more and more global in their business approach and are now offering the same products and services across borders, which decreases the importance of the sixth attention point.

---

* It is not certain that the difference in price between a link that does not and a link that does cross international borders will remain applicable, but for the foreseeable future this is assumed applicable.

Liebmann's list shows a series of attention points that is a selection of a series of aspects of management of international networks. As only a limited number of aspects can be studied here, a selection must be made. In an explorative case to be covered in Chapter 6, a longer list of international particularities will be found and examined. For the case study here, offering of telecommunications services is chosen, as well as regulatory environment.

Trends as a Basis for a Study

Trends, as addressed in section 1.2, also help in the choice of a case study. The first trend, "increased scale and globalization of networks," directly implies that the subject of international networks is increasing in importance. The second trend, the "changing regulatory environment," is considered in the literature to influence, for instance, the price of telecommunications services [vanCuilenburg 1995] and may therefore also influence cost-effective management of a network. Also, the magnitude of the trend is so important, with 69 countries going through drastic changes between 1998 and 2002, that the "changing regulatory environment" trend is a candidate for the case study. The regulatory environment is an aspect that is particular to a country, as it is directly related to applicable laws, which are often made by a country government or a regulatory body of the country. As a result, we will use the regulatory environment as an aspect for our study and will refer to it as regulatory environment. The third trend, "Telecommunications Services Offering" is a factor that is changing rapidly as the choice is widening, which confirms the importance of this aspect. The fourth trend, which concerns outsourcing, implies that international networks are more and more going to be managed by third parties, which confirms the importance of increasing the knowledge base on the subject of management of international networks.

Developing the Strategic Questions as a Basis for the Study

The first and the last strategic questions in section 1.3 directly concern the selection of a country for placement of a network. The second strategic question concerns the design of a network topology for an international network and therefore indirectly also requires a choice of countries among which to establish links.

With more than 200 countries in the world, the number of possible choices could be large and therefore the strategic questions should be approached in a more structured way. *Aspects* of countries should be considered that may give structure to making a choice. Individual characteristics of a country are also essential for the second strategic question on the design of an international network, which shows the need for knowledge of aspects of a country that influence the management of a border-crossing link.

This leads to the conclusion that one key strategic question fundamental to the above strategic questions can be formulated: *How do characteristics of a country influence cost-effective management of a network?* This question can form the basis for the case study.

To explore possible answers to the strategic questions, this case study is defined. Learning the answers can save international organizations time and money, make more effective use of resources, and provide more efficient communications among users of international networks around the world. The case study is therefore approached from the point of view of an international organization, rather than from that of the individual consumer, although the conclusions should be applicable to consumers as well.

In addition, a completely different reason to focus on international organizations and networks is that international organizations are more "progressive" [Grover 1995]. Grover found that international organizations have shown to be more open to adopting new technologies, such as Wide Area Networks, videoconferencing, and e-mail. When new technologies are used, what is learned from examination of these organizations can be even more valuable, given that they can act as a more useful learning experience for organizations that do not yet have the technology.

Two aspects for further study are the regulatory environment and the telecommunications-services offering.

Demarcation

We then need to divide up the case study to make the research manageable. For example, we will restrict the case study to certain types of international networks—in particular, private networks. Research on private networks may very well be applicable to public networks also, but this has yet to be verified. We do not use demarcation to restrict ourselves to a particular service or set of services. Data services, voice services, and other services may be provided by the network. Figure 1.2 shows the process of determination of the case study in a graphical format.

The case study will be approached with a model that we will refer to as the cost-effective management model. The cost-effective management model explains influences of the aspects of a regulatory environment and telecommunications-services offering (which we will now call "input parameters"

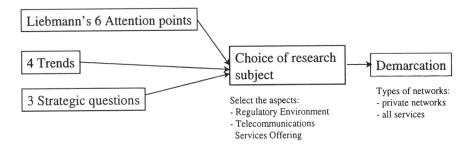

FIGURE 1.2 Process of determination of the case study.

of the model) on the cost-effective management, (which we will call "output parameter" of the international network). A country can be "scored" on its regulatory environment and telecommunications-services offering. The study will then assess if the scores have a relationship with cost-effective management of a network in the countries. The described flow of thoughts can be formalized in a research question.

1.5 RESEARCH QUESTION (WHAT?)

Having chosen the two input parameters, we consider their relationship with cost-effective management.

The research aims to show possible relationships between the *regulatory environment* and the *telecommunications-services offering* in a country on one hand and the *cost-effective management of the network* on the other.

Key words are printed in italics and will further be defined in the book, but can be generally explained as follows:

- The *regulatory environment* is the set of applicable laws, regulations and jurisprudence regarding telecommunications that is applicable in a country.
- The *telecommunications-services offering* is the set of available telecommunications services in a country.
- The *cost-effective management* is the degree to which the performance of the network fulfills the requirements of the users.

In order to clarify the "cause and effect" to be researched, the terms "input parameters" and "output parameter" can be used in the research question: *What heuristic relationships can be found between input parameters "regulatory environment" and "telecommunications-services offering" of a country in which part of the network operates, and the output parameter cost-effective management of the network?*

A note on types of relationships

Both the input parameters and the output parameter need to be quantified in order to alleviate comparisons between cases, however, the type of relationships that are sought were chosen to be qualitative rather than quantitative. For example, this means that a relationship should just show that there is a positive or inverse relationship between an element of an input parameter price of telecommunications services and cost-effective management, but does not have to show that if input parameter price increases by x%, the cost-effective management will decrease by y%.

Also, we are looking for a heuristic relationship, rather than a causal relationship. In a causal relationship, a proof of the relationship is necessary. When searching for a heuristic relationship, cases can be used to support the confidence in the relationship [Yin 1984].

1.6   RESEARCH APPROACH (HOW?)

Research methodology—the use of cases

As stated in the literature [Yin 1984], "a case study strategy is extremely useful for appraising a situation if the boundaries of the phenomenon are not clearly evident at the outset of the research and no experimental control or manipulation is used." Therefore, case studies in combination with literature studies are used as major sources of information. In addition, the experience in the international business of a major telecommunications operator creates ideas that are used in the project.

Case studies are used in *explorative case* form and in *test case* form. The difference between the two forms is that the explorative case is of much shorter duration than the test case and it does not test models or hypotheses for their validity. The explorative case's purpose is to examine an existing situation and generate ideas. The test case is a more lengthy procedure that applies models that have been developed and has as its main purpose to test and explain the relationships expressed in the propositions that the models suggest, and create ideas for improvement of the models.

Relationships in the case studies

In the case studies, a part of the network that can be studied is first chosen. In order to study relationships, the regulatory environment and the telecommunications-services offering are scored for each country in which part of the network being studied is operational. Then statistical analysis is done to get insight in the existence and type of relationships between the input parameters and the *cost-effectiveness of management* of the part of the network in that particular country.

Figure 1.3 depicts the relationships sought in the research question.

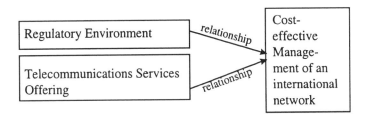

FIGURE 1.3 Graphical depiction of the relationships sought in the research question.

Research Methodology: 10 Steps

1. Literature study and research question

Step 1 in the research methodology has the purpose of clearly defining the case study using literature and experience in management of networks and in international aspects, so the research question can be formulated. The outcome of step 1 is the research question, with a research methodology that clearly describes how the research methodology can be answered. Chapter 1 shows an overview of the case study and the choice of the research question.

2. Development of the theory for input parameters and the output parameter

With the research question and the case study, several theories are studied and their value with respect to answering the research question is assessed. Even when a theory seems to have relatively little value toward answering the research question, a summary is made for further reference, as ideas that are used in the theory may be used in other steps of the research. Special focus is given to the theories on network management in general and on international aspects of countries. Chapter 2 shows the result of this development of theories.

3. Development of the input parameter regulatory environment

Using the existing literature found in steps 1 and 2, the input parameters can be described in a systematic and quantitative way. An analysis should be done of what elements constitute a regulatory environment. A model should be developed to express the input parameter regulatory environment of a country in a quantitative way. The model can initially use starting values (values chosen using experience or estimates) that can later be fine tuned. The result of the development of the regulatory environment is shown in Chapter 3.

4. Development of the input parameter telecommunications-services offering

Different elements of a telecommunications-services offering will be determined. The elements that are to be included in the model that describes the offering should be selected after analyzing the level of importance of the various elements for international organizations by means of a survey. The result of this step, a model that describes the regulatory environment, is depicted in Chapter 4.

5. Development of the output parameter cost-effective management

After both the input parameters have been modeled and can therefore be quantified, the output parameter, cost-effective management, needs to be researched on how it can be measured and quantified. For this research, it is essential to find a method of quantification that is both accurate enough to determine the existence of relationships as mentioned in the research question, and easy to use. If possible, cost-effective management should be expressed in one formula that comprises certain parameters of a network and its management. Chapter 5 shows the chosen formula to measure cost-effective management.

6. Explorative case

The explorative case is used to examine a situation and generate ideas [Yin 1984] and in this research primarily to examine the situation of applying the models that describe the two input parameters and the output parameter. The practical environment of the explorative case may show possible complications with the application of the models of the input and output parameters and can teach us about the circumstances in which the models should be used.

Second, however not as a direct contribution to the solution of the research question, the explorative case can be used to assess other experiences of management of international networks. These experiences may form a basis for a list of "international particularities" that should be considered for the management of international networks.

The first result of this step is a series of experiences in the explorative case that can help refine or better apply the models of the input parameters and the output parameter. The second result is a list of several international particularities that are found in a practical situation of operating an international network. The explorative case is described in Chapter 6.

7. Development of a "cost-effective management model"

As steps 3 through 5 developed the input parameters, which are input to a model, and the output parameter, which is the output of a model, we now consider the model itself. In addition, step 6 helped obtain practical experience with input parameters and output parameter. Step 7 seeks to describe the cost-effective management model. Furthermore, literature studied in step 2 may be used.

The cost-effective management model gives an assessment of a relationship between the input parameters and the output parameter by stating propositions. Propositions are statements on the relationships, which can be tested with example cases. The combination of propositions should embody an answer to the research question, which makes the cost-effective management model of prime importance in the case study. Where possible, propositions should be explained with the help of literature or experience from earlier steps.

The result of this step is a series of propositions on relationships between the input parameters and the output parameter, as well as an explanation of why these propositions could be valid. Development of a cost-effective management model is described in Chapter 7.

8. Two test cases

After the cost-effective management model is developed, two *test cases* should be carried out in international organizations. The main purpose of the test cases is to apply the cost-effective management model in practical situations and test the propositions that were developed in step 4. Two case studies in international organizations with international networks should be identified with the following requirements:

- Each case concerns an organization that operates in a different industry.
- Each case has a considerable number of countries, so that the statistical analysis can be done with confidence. A minimum number of five countries is assumed for our cases.
- Each case has a stable network environment that does not change drastically within the time frame of observation.
- Each case operates in countries where data on the input parameters regulatory environment and telecommunications-services offering is available.

Due to the nature of the case method, the test cases cannot *prove* the propositions of step 7 [Yin 1984]. They can only strengthen or weaken confidence that a proposition is valid and the confidence that an actual relationship between input parameters and the output parameter exists.

As mentioned in section 1.4, there is no intention to show a *quantitative relationship* between the input parameters and output parameter, but rather a *qualitative relationship*.

The result of this step is a choice of two organizations where cases are to be done, a detailed description of the case situations and application of the models that score the input parameters and the output parameter for the networks in the case situations. The two test cases are described in Chapter 8.

9. Statistical analysis of the test cases

In this step, statistical analysis of the test cases is done with the quantitative representation of the input parameters and the output parameter. The use of statistical analysis may enable us to test the validity of the propositions and reveal more of the nature of the relations between the input parameters and the output parameter.

The statistical analysis method to be used is regression analysis. Regression analysis is a method for showing a relationship between two series of data, in this case the input parameters and the output parameter. Both so-called *simple regression*, which involves one input parameter, and *multiple regression*, which involves multiple input parameters, can be used.

Regression analysis should be done for each of the input parameters, but it should also be done for individual elements of the input parameters, as this can reveal more detail on possible relationships that may exist just between elements of the input parameters and the output parameter. A limited series of multiple regressions should also be done to assess the potential for relationships between a combination of multiple input parameters and the output parameter. Multiple performance indicators are performed to adequately measure the performance of a network. In that case, a regression analysis is done for each of the performance indicators separately and multiple indicators of cost-effective management are calculated as a result, which means that multiple output parameters are used.

The result of this step is a series of statistical indicators, including R-square and correlation, that show the likelihood of a relationship between each of the input parameters and the output parameter. This step is depicted in Chapter 9.

10. **Conclusion and further research**

Using the cost-effective management model of step 7 and the statistical indicators of the test cases of step 9, conclusions should be drawn on the validity of the propositions. Furthermore, potential refinements for the cost-effective management model or for the models describing the input parameters and output parameter can be suggested.

The results of this step are an assessment of the validity of the propositions, a suggestion for possible refinements, and a proposal for future research. Step 10 is described in Chapter 10. Figure 1.4 shows an overview of the research methodology with all steps in one figure.

1.7   NETWORK MANAGEMENT: CURRENT LITERATURE

Theories for management of international networks and for analyzing telecommunications services and regulatory environments exist in various areas of current literature. Most common areas are *management science, legal and regulatory sciences, computer science* and, to a lesser extent, *organizational behavior*. The current literature mentioned here is only a short list that is the basis for the theories to be developed. Overviews of literature of specific areas of network management can be found in [Verdonck 1992, Terplan 1992, and vanHemmen 1997].

1.7.1 SERVICE LEVEL AGREEMENTS

A service level agreement (SLA) is an agreement between a network and the organization using the network. It covers the management of the network, including level of performance. SLAs are not always in place in organizations and may have various forms, such as a written agreement, written departmental

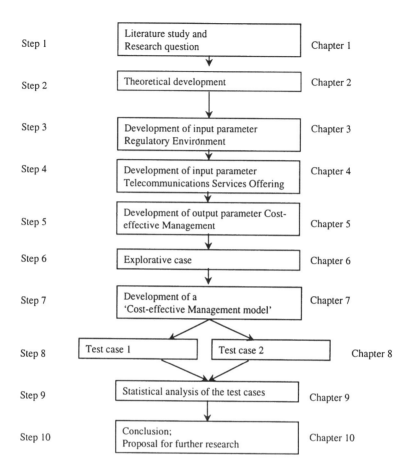

Step 1   Literature study and   Chapter 1
              Research question

Step 2   Theoretical development   Chapter 2

Step 3   Development of input parameter   Chapter 3
              Regulatory Environment

Step 4   Development of input parameter   Chapter 4
              Telecommunications Services Offering

Step 5   Development of output parameter Cost-   Chapter 5
              effective Management

Step 6   Explorative case   Chapter 6

Step 7   Development of a   Chapter 7
              'Cost-effective Management model'

Step 8   Test case 1   Test case 2   Chapter 8

Step 9   Statistical analysis of the test cases   Chapter 9

Step 10   Conclusion;   Chapter 10
              Proposal for further research

FIGURE 1.4 Research methodology.

objectives, or verbal agreements [deLooff 1996, Passmore 1996, Ananthanpillai 1997].

1.7.2 MANAGEMENT CONTROL AND MAINTENANCE PARADIGM AND RELATED THEORIES

The management control and maintenance paradigm (MCM paradigm) is the basis for a series of theories that help establish insight in management of international networks [Looijen 1998]. The MCM paradigm and its use in this research can be understood by starting with the "value chain model."

The "value chain model," as developed by Porter [1998] gives the opportunity to look at an international network as an entity that uses (purchases) and delivers (performs) services. An international network may add value to services in many different ways. For instance, it can deliver the same tele-communications services as it purchases, but with a different kind of billing, or it may even bring services across borders, add customer care services* or help desks, or add a "single-point-of-contact" with which the users can interface when there are problems with the network [Bishop 1996].

The *international network* as depicted in Figure 1.5 uses certain tele-communications services as its input and delivers certain services as its output. The reason for using this figure is that it forms a clear way to differentiate services performed for the users of the network and services that are purchased from other parties. It thus allows us to describe the network in terms of the value that it adds to purchased services.

FIGURE 1.5 Telecommunications services in the international network value chain.

The network management paradigm [Van Hemmen 1997], shown in Figure 1.6, was developed out of the MCM paradigm [Looijen 1998]. A "network system" is defined as the combination of network, network ser-vices, characteristics, users, requirements, preconditions, and procedures. An international network system is a network system with a link that crosses an international border. When we refer to an "international network," actually an "international network system" is meant. In other words: the elements of the network system, such as network services, characteristics, users, require-ments, preconditions, and procedures are considered included.

Arrows A and C, in Figure 1.6 show that the network management paradigm embodies the value chain as depicted in Figure 1.5.

The services denoted by the arrows A, B, and C are generally categorized as telecommunications services. The service value model, to be described in section 2.3, will categorize them in more detail. Examples of influences

---

* Customer care services are the services that provide information to support the use of the telecommunications services. Customer care services can be provided by people, who assist the users with the use of the services and payment for them, but there are also other ways to provide customer service, such as the use of web sites or voice-response phone systems.

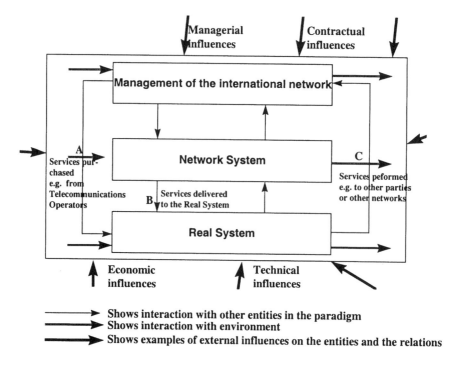

FIGURE 1.6 The network management paradigm.

from the outside world on the international network and its management are various and are categorized by Looijen [Looijen 1998] as managerial, contractual, economic, and technical.

1.7.3 OSI NETWORK MANAGEMENT FRAMEWORK

A commonly used model for describing activities for management of networks is the OSI network management framework [Terplan1992, Obsitnik1994, Ananthanpillai1997]. The OSI network management framework identifies five "functional areas."

1. Configuration management
2. Fault management
3. Performance management
4. Security management
5. Accounting management

A short description of the functional areas follows (see also [Terplan1992]).

Configuration management controls physical, electrical, and logical inventories, maintains vendor files and trouble tickets, supports provisioning and order processing, defines and supervises SLAs' ongoing changes, and distributes software.

Fault management is a set of activities required to dynamically maintain the network service levels, such as pinpointing problems and performance degradation quickly, and, when necessary, initiating controlling functions, which may include diagnosis, repair, testing, recovery, and backup.

Performance management is required to continuously evaluate the performance of network operation, to verify how service levels are maintained, to identify actual and potential bottlenecks, and to establish and report on trends for management decision making and planning.

Security management is a set of activities for the ongoing protection of the network and its components, such as the protection against unauthorized entry of the network, accessing an application, and transferring information in a network.

Accounting management is a set of activities for collecting, interpreting, and reporting costing- and charging-oriented information on resource usage, such as processing of accounting records, bill verification, and carrying out charge-back procedures.

The functional areas of the OSI network management framework can be used in various ways. In this book, the use is *categorization* of activities for network management, which will be used in the case situations. With a proper categorization of activities, organization diagrams of case organizations can be better understood and international aspects can be categorized in a standard way that can be recognized by people not otherwise familiar with the situation.

1.8 SUMMARY

Managing an international network can consume as much as two thirds of the total budget for communications. Since an international network crosses borders and operates in various countries, aspects of the countries, such as the language, telecommunications operators, the offering of telecommunications services, and the telecommunications regulations may influence cost-effective management of the network.

This book focuses on modeling two *aspects* called *input parameters*, which may influence the management of an international network: the "regulatory environment" and the "telecommunications-services offering" in the countries in which the network is located. The research should develop a model (the cost-effective management model) that explains influences of the two input parameters on an output parameter.

The cost-effective management model is intended to "support" the strategic qustions of international organizations, in particular related to the influence of country aspects, but the cost-effective management model also gives insight into international organizations and governments as to what the influence of regulatory environment and telecommunications-services offering is on cost-effective management of a network, so governments can adjust their policies to improve their country's attractiveness for international networks. Chapter 2 will develop a series of theories that can be used to further research management of international networks.

2 Theoretical Development

In this chapter, various theories are further developed for management of international networks, as required by step 2 of the research methodology. Theories for management of networks in general are discussed in section 2.1, for the input parameter regulatory environment in section 2.2, for the input parameter telecommunications services offering in section 2.3, and for the output parameter cost-effective management in section 2.4. Theories for crossing international borders are addressed in section 2.5.

2.1 TELECOMMUNICATIONS MANAGEMENT NETWORK: GENERAL THEORIES

Telecommunications management network (TMN) is a standard that supports the management of telecommunications networks [Verdonck 1992, van Hemmen 1997]. The International Telecommunications Union, section Telecommunications Standardization, (ITU-T) has been developing this standard during the 1990s. The architecture of the standard was complete as of 1998, but several details are still to be filled in.*

The TMN architecture is divided into four layers, according to the objects that are managed. They are:

1. Network-element management
2. Network management
3. Service management
4. Business management

Network-element management
The network-element management layer takes care of the management of a certain class of network elements, such as network elements of a particular vendor. The class of equipment may need to be controlled by a specific network management system. This layer guarantees that the network management information will be translated into standards that are used in the network management layer.

---

* Recommendations in M.3010 "Principles for a TMN"

Network management

The network management layer, with its overview of the network as well as the individual elements, manages the total network. MCM of network elements is now integrated and vendor-specific issues are already taken care of in the network-element management layer.

Service management

MCM is carried out for specific services that are running on the network. Management at this layer makes it possible to directly support network users who use a particular service. The functions of this layer can be divided in two groups:

1. Functions that support operational customer processes
2. Functions that support service managers, that are responsible for the financial results of a certain service including sales, billing, service quality, and capacity planning

Business management

Business management is a layer that focuses primarily on the business associated with the services, including their cost effectiveness. The business management includes the management of the network at a strategic level with longer-term trends, the planning of new services, security, and quality.

2.2 THEORIES FOR INPUT PARAMETER REGULATORY ENVIRONMENT

The literature has presented a few theories that show a categorization of the regulatory environment in a country. A theory found, for instance, is the Meta Telecom Maturity Model. The most detailed theory found is the categorization by the Yankee Group, called "regulatory index" in their reports. Another model is the "liberalization index" used by the OECD. Furthermore, there are various treaties and laws that have influence on a large number of countries. Those treaties and laws will also be addressed in this section.

Meta Telecom Maturity Model

This theory, developed by the Meta Group [Johnson 1997] is based on a checklist against which a country is evaluated. Using the checklist, which contains regulatory and infrastructure aspects of a country, the Meta Group classifies countries as Class A (advanced), Class B (highly developed), Class C (developed) or Class D (growth). The items in the checklist are:

- Privatization: Is the telecommunications operator privatized?
- Independent regulation: Is the regulatory body affiliated with the main telecommunications operator?
- Foreign competition and ownership: Is investment in telecommunications operators allowed by foreign entities?
- Competitive interconnection and resale: Is connecting to the main carrier's network permitted?
- Alternative infrastructure: Are multiple infrastructures for voice, data, and wireless services available?
- Universal service: Is universal service a mandatory offering?
- Enhanced services: Are telecommunications operators offering advanced services?

The categorization in classes is then roughly defined by checking what questions are answered affirmatively. If the first three answers are "yes" and the rest of the conditions are at least 90% met, the country can be placed in Class A. If a country fails one of the first three conditions it is placed in Class B or below. The differences between B and C classifications are mainly in the offering of infrastructure. Class B (e.g., Germany, Australia) offers ubiquitous infrastructure, but Class C (e.g., Malaysia) offers only infrastructure in major business areas. Class D shows the least regulatory freedom and infrastructure presence.

The Meta Telecom Maturity Model is fairly coarse and does not show detailed criteria for placing the countries in the classes, but does, at least, give insight into what conditions should be satisfied for a deregulated country in the opinion of the Meta Group. The model was developed for the purpose of making network managers aware of the different environments they might encounter when doing business in different countries, and to sensitize them to the need for different tactics when negotiating service agreements with their suppliers.

Yankee Group Regulatory Index

Another theory that is available in the literature is the "regulatory index" of the Yankee Group. This index is simply a matrix with services and a "score." The score is expressed in a number between 1 and 5, where 1 indicates that regulations are highly restrictive, or that no reform has been implemented; a 5 indicates that regulations are very liberal, or that radical reform has been implemented [Yankee 1996].

There is no strict assignment of the numbers to a specific regulatory situation, but many examples are given in the Yankee Group reports.

Table 2.1 gives an example of the use of the Yankee Group regulatory index for the country Japan. The left column shows various "markets" (as they are called in the Yankee Group literature). For each of the markets, a number is assigned to show the score on the 1–5 scale, but comments are added in almost every row to show that the regulatory index can be different for different services, or the comment is used just in general as a more detailed description.

TABLE 2.1
Example of a Regulatory Index Table on Japan

Market	Score and Comments
Terminals	5: Terminal equipment markets for both fixed and wireless (mobile phone, cellular) services have been liberalized
Basic Services	4: Basic services using the public network can be provided only by so called "Type 1 carriers"
Value Added Services	5: The market is open
Independent regulator	3: Ministry of Post and Telecommunications" (MPT) regulations (or wishes) have been generally vague, unwritten. However, recent communications by the MPT indicate that it will clarify rules and regulations pertaining to telecommunications.
Private network restrictions	4: Private networks are allowed
Mobile services	4: Currently pricing needs to be approved by MPT. According to recent announcements, pricing will be liberalized in 1997
Satellite services	2: Still heavily regulated

Source: Yankee Group, 1996

This regulatory index is not considered an adequate theory for our study, as its scoring is very unclear and it mixes service categories with other aspects, such as the regulatory body. The title "market" is very confusing; we would rather call it "services categories," and the entries "independent regulator" and "private network restrictions" would fit better in a separate table.

OECD Liberalization Index

The Organization of Economic Cooperation and Development (OECD) liberalization index [OECD 1997] is comparable to the Yankee Group regulatory index. It also takes groups of telecommunications services in a country and lists the state of competition in the market for those services. An overview of the different values of the OECD liberalization index is shown in Table 2.2. The OECD uses letter codes to describe the state of competition in different kinds of services. Table 2.2 shows the letter codes in the first column. They designate a state of competition described in the second column and a liberalization index, stated in the third column. The 1993 report [OECD 1993] gives an explanation of the codes and connects the index numbers with the codes. The later reports [OECD 1995, 1997] use only the codes, not the index numbers.

TABLE 2.2
OECD Liberalization Index

Code	Description	Liberalization index
C	Competition	2
PC	Partial competition (in a particular geographic area)	1.5
D	Duopoly	1
RD	Regionalized duopoly	1
B	Competition allowed at the border of concessions	0.5
M	Monopoly	0
N	No service	0

The OECD adds up the scores per country for a set of eight different services:

1. Public Switched Telephone Network (PSTN) local within a city or region
2. PSTN domestic long-distance
3. PSTN international
4. Packet Switched Data Network (PSDN)
6. Mobile analogue voice communications
7. Mobile digital communications
8. Mobile paging

The resulting scores range from 0–16, with a higher number denoting a more competitive environment.

The OECD model does not apply much detail and the different states of competition are not described with criteria, but it is the most adequate model as it is one of the more important building blocks for our regulatory environment model, which is to be developed in future chapters.

Concluding on theories for the regulatory environment, we assess that the theories found are not sufficiently detailed to function as input parameter for the relationships mentioned in the research question and they will therefore be further developed in Chapter 3.

2.3 THEORIES FOR INPUT PARAMETER TELECOMMUNICATIONS SERVICES OFFERING

We developed a theory, named *service value model*, to get an overview of and categorize the communications services. The communications services are delivered, or "performed," by a network (arrows B and C in Figure 1.6) and used by a network (arrow A in that same figure). The model is based on the OSI (Open Standards Interconnection) model, developed by ISO (International Standards Organization).* The OSI model [Black 1993, Tanenbaum 1997] includes layers for which protocols for communication were developed by the industry and standards organizations. The model is somewhat similar to an existing model developed for mobile services by Arnbak [Arnbak 1997].

In our service value model, the layers are grouped so that interfaces between the groups of layers can be expressed in common standards as much as possible and so that it matches with the way most present telecommunications regulations in countries are structured. Services are classified in the service value model in *service value levels*.**

The model identifies three "service value levels" as shown in Figure 2.1.

A service value level is a group of services determined by the levels in the OSI model. As the highest layers of the OSI model use interfaces that are still in the process of being standardized, the boundaries of the service value levels may vary slightly, depending on the actual protocols and interfaces.

---

* Understanding the OSI model, is, however, not necessary for the comprehension of the essence of this book.

** "Service value level" should not be confused with a "service level" as commonly used in agreements to provide a set of services (service level agreements). The services meant in the service level agreement are usually a much broader set of services.

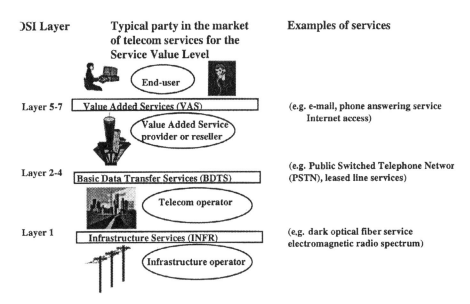

OSI Layer	Typical party in the market of telecom services for the Service Value Level	Examples of services
	End-user	
Layer 5-7	Value Added Services (VAS) / Value Added Service provider or reseller	(e.g. e-mail, phone answering service Internet access)
Layer 2-4	Basic Data Transfer Services (BDTS) / Telecom operator	(e.g. Public Switched Telephone Networ (PSTN), leased line services)
Layer 1	Infrastructure Services (INFR) / Infrastructure operator	(e.g. dark optical fiber service electromagnetic radio spectrum)

FIGURE 2.1 The service value model.

Value added services form the top layer of services and can be characterized as the services that are defined in protocols of OSI layers 5–7. They are services that the end-users actually use. In many countries, value added services are the first services where competition between suppliers is allowed and the supervision of regulatory bodies is only minimally present.

Basic data* transfer services are services that provide the transport of data (which can mean various kinds of information) in a certain format and fit in the OSI model in layers 2–4. Protocols for these services are usually standardized in an international environment to facilitate connection to equipment on the network. Examples are packet switched data transfer service, which transports information in information packets, and international leased lines, which are dedicated lines between points in the network that can transport a specified bandwidth of data.

Infrastructure services are services that consist of making available a physical connection with certain technical specifications between two locations. They fit in the OSI model in layer 1. In most countries, the provision of infrastructure services is still the exclusive right of the public telecommunications operator (PTO). There is not much standardization yet in infrastructure services, as they have not been generally available until recently.

---

* Data is placed between brackets since "data" is a general term. So no specific choice for fax, voice, computer data, or video is meant.

An example of an infrastructure service is *dark optical fiber service*, which offers a fiber connection with certain specifications between two points, where the user of the service gets access to the endpoints of a fiber in point A and point B.

2.4 THEORIES FOR OUTPUT PARAMETER COST-EFFECTIVE MANAGEMENT

Measuring the output parameter cost-effective management of the network apparently is an extensive task, and very few models have been found in research that actually calculate a quantitative value for cost effectiveness of the management of a network [Loe 1994, Terplan 1995]. Terplan has developed a method for *benchmarking network management*, which consists of four phases. The phases are:

1. **Data Collection**. Data on the management of the network is collected through various means.
2. **Data Consolidation**. Comparisons of data are made with best practices, standards, or industry averages.
3. **Gap Analysis**. Details of the gap between the results of the previous two phases for all functions of network management are analyzed.
4. **Elaborating the recommendations**. The gap identified in the previous phase become the basis for improvement or for supporting the decision to outsource certain network management activities.

Terplan's method and its phases are defined in detail and form an extensive basis to perform a thorough assessment of benchmarks. For the purpose of modeling the output parameter cost-effective management, only a relatively small part of Terplan's method will be applicable, as the purpose is not to compare cost-effective management of a network with industry standards and to do detailed analysis, but merely to get a high-level overview with some kind of quantification of the cost-effective management of an international network in different countries that will result in numbers comparable among countries.

2.5 CROSSING INTERNATIONAL BORDERS

Finally, we examine a particularity that every international network has: the *international link(s)*. The management of the international link will be discussed first, then tools for support of management of the international links will be addressed. Both subjects are examined from a theoretical background. Practical experiences with this and other international particularities will be mentioned when the explorative case is discussed in section 6.2.

2.5.1 MANAGEMENT OF LINKS

Using the service value model, the management of a link may be carried out by different parties, e.g., telecom operators, such as PTOs, for certain value levels of the link. In Figure 2.2, an international link between nodes in countries A and B is a link that delivers basic data transfer services to the international organization, which uses its own information systems to provide value added services to its customers.

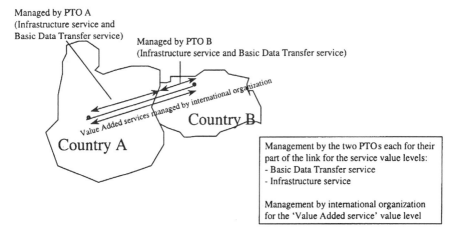

FIGURE 2.2 Example of management of a link that crosses an international border.

When two different telecom operators manage the link for the two service value levels, that has important implications for the management of the total network. An example is fault management, which requires that, for example, if a user of a service calls the help desk, there will be a quick answer as to when the problem will be solved and appropriate action will be taken to solve the problem. If the query is at the *basic data transfer service* or *infrastructure service* value level, it may be possible that several PTOs will need to be contacted before the problem is found. Dealing with the PTOs requires special procedures and possibly special clauses in the SLAs between the PTOs and the international organization to control these problems and to maintain an adequate level of support to the end-user [Passmore 1996, Ananthanpillai 1997].

The need for special procedures or clauses is illustrated by the following example: An international organization with several border-crossing links has SLAs with PTOs to restrict outage times for the links to a predefined minimum over a time period. As a backup scenario, the organization might have SLAs with other telecom operators to fulfill similar links between similar nodes. The other telecom operators should have a clause in their contract that specifies a different PTO for its infrastructure services from the one with which the organization already has a contract. The extra clause may avoid both links being physically being routed over the same infrastructure and therefore possibly failing at exactly the same time.

2.5.2 IMPROVING MANAGEMENT OF LINKS

Management of an international network can be supported by "tools." As the presence of an international link, which is a link that crosses an international border, is the key difference between an international network and a noninternational network, tools that help manage international links are of particular importance.

Analysis of the principles of international networks and experiences mentioned by users in surveys, such as a survey by DataComm [Heywood 1997] resulted in the development of the responsibilities matrix. In addition to this matrix, two other tools are mentioned: *one-stop shopping* and the *core skills* model.

Responsibilities Matrix

One tool developed in a research project is the *responsibilities matrix*. The responsibilities matrix (see Table 2.3) can serve as a tool to check that all responsibilities for management covered in the OSI management framework have been taken care of. All cells in the matrix need to be filled with the names responsible for particular functional areas to ensure that all responsibilities for management of an international link are taken care of. The parties mentioned in a cell together must be responsible for the *complete* management of the service value level in that functional area.

Writing down responsibilities has proven to be of value, even in relatively simple cases with few international links. It can, for instance, help focus PTO A during the negotiations with the organization on what it will have to work out with PTO B or other service providers or operators to provide full coverage of particular functional areas for each layer in the service value model.

TABLE 2.3
Example of a Responsibilities Matrix

Service Value Level	Configuration management	Parties responsible for the functional areas Fault management	Performance management	Security management	Accounting management
Value Added Service	Int. Org./ Technical dept.	Int. Org/ Consult dept.	Int. Org/ Technical dept.	PTO A	Int. Org / Security dept.
Basic Data Transfer Service	PTO A (part) PTO B (part)	PTO A (part) PTO B (part)	PTO A (part) PTO B (part)	PTO A (part) PTO B (part)	PTO A (part) PTO B (part)
Infrastructure Service	PTO A (part) PTO B (part)	PTO A (part) PTO B (part)	PTO A (part) PTO B (part)	PTO A (part) PTO B (part)	PTO A (part) PTO B (part)

One-Stop Shopping

One-stop shopping is a term generally to describe the availability of various different services and products from one supplier, instead of having to use multiple suppliers. It may not look like a "tool" at first glance, but a supplier that offers one-stop shopping is seen as offering a tool to its client [Tice 1997]. The supplier may offer it as an option that requires separate payment. In the case of one-stop shopping, one agreement between the telecom operator and the international organization should be sufficient. For example, if PTO A implements the one-stop-shopping concept for basic data transfer services [Elixmann 1996], there might be an SLA between the international organization and a PTO that authorizes the one-stop shopping concept in country A to maintain the service level on that link. PTO A has an SLA (or other type of contract) with PTO B for the part of the link in country B. For all functional areas of management of the link, the SLA should cover what activities either party should perform and how the performance, according to the SLA, can be verified.

The core skills model

The "core skills" model, the third tool to be addressed here, was found in the literature [Allott 1997]. The core skills model shows that a telecom operator can manage the network best by:

- Buying the right network
- Running the network well
- Filling up the capacity on the network
- Competing effectively

An explanation that Allott shows is:

- "Buying the right network" is interpreted as buying a network that exactly fits the needs in terms of types of services.
- "Running the network well" is a very ill defined term, that, according to Allott, includes "remote management and managed network service provision" and the use of new management for new technologies, such as the combination of voice and data in technologies like ATM. Whether a network is "run well" is mostly determined by what the users indicate in their SLAs.

- "Filling up the capacity on the network" is developed with the analogy of other industries, such as the airline industry, in mind. These industries also have very high fixed costs and relatively low variable costs. They show the "utilization factor" as a prime indicator of their efficiency. As an illustrative example, past financial results have confirmed that airlines' numbers show a positive correlation between utilization factor and financial results.

"Competing effectively" is considered the result of managing SLAs and the ability to customize management for certain customer requirements.

We considered the core skills model to be more focused on getting financial results and competing in the services market as a telecom operator, than on improving the management of the international network as such. We did not find an immediate application for them in this study, although some of the conclusions of the core skills model might be useful for analyzing the networks in the case studies.

2.6 LEGAL IMPLICATIONS OF CROSSING INTERNATIONAL BORDERS

A link that crosses an international border also has legal implications, such as the impact of import and export laws and tax laws in countries on both sides of the link. Data that flows across a border (also called transborder dataflow) can be subject to taxes for import or export. An example of a situation where taxes are important is when software is developed in a low-wage country and shipped to another country for sales. Some data may not be exported, for example confidential governmental data. Also, some data may not legally be imported, such as pornographic material in several Middle-Eastern countries. We just note these legal implications of crossing international borders, but will not further research them. In the explorative case, we will search for various "international particularities" and will list the ones we find, among which may be these legal implications as well.

2.7 SUMMARY

Although some theories have been developed for management of networks, literature on management of *international* networks is scarce. This chapter shows a combination of existing theories like *Meta Telecom Maturity model* and the *core skills model*, which are adapted from the *MCM Paradigm*, and self-developed theories such as *management of links* with a responsibilities matrix and the *service value model* that can be used to model specific areas of the research subject.

The available tools to model the regulatory environment and the telecommunications services offering in a country are not sufficiently detailed for immediate use in our research. The regulatory environment and telecommunications services offering will therefore be further developed in Chapters 3 and 4.

Cost-effective management is also an area with relatively underdeveloped tools, although, on the subject of benchmarking, detailed theories by Terplan [Terplan 1995] are available. Cost-effective management will be further modeled in Chapter 5.

3 Regulatory Environment

Step 3 in the research methodology develops a model for the input parameter regulatory environment, which is covered in this chapter. The regulatory environment is one of the two aspects of the research question and should therefore be described so that the description can be used as an input parameter for the cost-effective management model.

The regulatory environment can consist of various types of regulations that influence international networks, including regulations concerning secrecy of information flowing across international borders, privacy of citizens whose information is transported across international borders, and regulations that concern parties that may compete in offering telecommunications services.

In this book, we describe the regulatory environment for a *country*, as is common in most literature. So, although particular reference may be made to regulation for a different entity, such as European Union or a particular province, the entity for examination of the regulatory environment will be the country.

Literature [OECD 1997] suggests that the regulatory environment plays a role in the management of a network. Through research on the influence of the presence of competition (which we will define as one element of the regulatory environment) on the price of telecommunications services, the OECD has found a relationship*. The regulatory environment that we propose is divided into three elements: The legal framework for competition, shown in section 3.1, the regulatory body, shown section 3.2, and competition that is actually active, depicted in section 3.3. The first two elements are prescriptive and the third is descriptive. Furthermore, section 3.4 gives an overview of the regulatory environment as a whole, and section 3.5 shows new developments concerning the regulatory environment that are taking place.

3.1 LEGAL FRAMEWORK FOR COMPETITION

The first element we chose for modeling the regulatory environment is the legal framework for competition. For our purposes, the legal framework for

---

* Van Cuilenburg and Slaa [1995] have done research on the statistical correlation between liberalization and economic factors, in particular innovation (the introduction of new products or services). Their research shows a positive correlation between liberalization and innovation. This will be used in Chapter 7, where propositions are developed and explained.

competition of a country consists of a series of laws regarding telecommunications competition. A country's applicable telecommunications laws can be assessed in terms of the possibility they create for telecommunications services suppliers to compete when offering their services. Laws may be enacted at the country level (e.g., by the country government) or at a higher level, the so-called supranational level [Korthals Altes 1999], or at a lower level in some countries (e.g., by a state or province). As we have taken the country as the unit of analysis, we look at applicable laws in the country. However, the applicable laws in a country may have been made by supranational organizations such as the European Union (EU). Laws from the supranational organizations may have direct effect on the citizens and companies in a country. For instance, in the case of the EU, laws or other regulations only have direct effect if they are "self-executing," which is usually defined in the laws or regulations themselves. An essential EU directive on the road to full competition in the European Union was the Open Network Provision (ONP) directive.* The ONP directive mandated the opening to other operators of infrastructures that are controlled by a dominant or monopoly operator. The World Trade Organization (WTO) is another example of a supranational organization. The WTO, however, does not make laws, but forms agreements that are ratified by country governments that then implement those commitments in laws. When we keep the unit of analysis the country, a mix of different laws can be applicable. In practice, national governments ensure that laws at a country level do not contradict laws that have influence on that country at a supranational level. In Figure 3.1, an example of different legal frameworks is shown. In the example, the country is assumed to be within the EU.

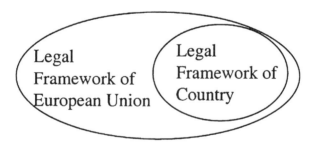

FIGURE 3.1 In a specific country, laws from different legal frameworks may be applicable.

* "Framework directive for ONP," established at the meeting of the Commission of the European Community, 28 June 1990, further detailed in meetings in 1992 and 1993 for four services: leased lines, packet switched data network, ISDN and PSTN.

It is essential to know what government or body enacts a particular law or regulation, as this has a strong influence on the methods for its enforcement. The legal framework for competition as we describe it is represented by a listing of telecommunications services categorized in service value levels (see section 2.3) in the country, and symbols showing for each level the laws rule that that service value level is open to more than one supplier.

Service value levels are part of the service value model, a model that we developed to categorize telecommunications services according to their place in the value chain for the provision of telecommunications services. Laws sometimes differentiate between types of services. The legal framework for competition, however, does not address such differences between types of services. It addresses only differences between service value levels of the services.

The legal framework for competition is described here in terms of its actual state. Multiple aspects, such as political, social, and economical aspects, may influence it.

Four categories are used to show the content of the laws with respect to allowing suppliers to offer the telecommunications service. The categories used are: "M" (monopoly), "D" (duopoly), "PC" (partial competition) and "C" (competition).*

M means that only one supplier is allowed to supply telecommunications services of that service value level; D means two suppliers are allowed to compete in supplying services of that service value level. PC means that there is only competition in certain areas. C means that there are several suppliers of similar services and that there is competition in the whole country. Current literature [OECD 1995] considers the order of liberalization from less liberalized to more liberalized as: M, D, PC, C.

To use the regulatory environment in a quantitative way, we would like to score the categories, with a higher score meaning a legal framework for competition that allows more freedom for parties to offer telecommunications services. We have chosen starting values for the scores as an assumption, using personal experience and the literature: M, D, PC and C are awarded 0, 1, 2, and 3 points respectively as starting values, although some practical experience asserts that the difference between M and D is larger than between the other categories. This results in the starting values as shown in Table 3.1.

---

* These categories are also used by the Organization for Economic Cooperation and Development [OECD 1997].

TABLE 3.1
Starting Values for the Legal Framework
for Competition

legal framework for competition	Score Starting value
M	0 points
D	1 point
PC	2 points
C	3 points

These starting value scores are used to quantify the legal framework for competition and do statistical analysis in future case studies.

The legal framework for competition is a *prescriptive* element, which means that it quantifies a situation that is prescribed. The use of laws, for instance, can prescribe a behavior, but does not give any feedback if the behavior actually occurs. Table 3.2 gives an example of scores for the legal framework for competition for sample countries X, Y, and Z.

TABLE 3.2
Example of a Legal Framework for Competition for Sample
Countries X, Y, and Z

	Legal framework for competition						
	Value Added Services		Basic Data Transfer Services		Infrastructure Services		TOTAL legal framework for competition
	Code	Score	Code	Score	Code	Score	Total score
Country X	C	3	M	0	M	0	3
Country Y	C	3	C	3	C	3	9
Country Z	C	3	C	3	C	3	9

3.2 REGULATORY BODY

The second element of the regulatory environment is the Regulatory Body, written with capital letters as the representation of the regulatory body in the input parameter is meant. Regulatory Bodies are organizations that are responsible for implementation of part of the regulation and, in some cases, are responsible for development of regulation. In EU legislation, they are also called National Regulatory Authorities (NRAs). Regulatory bodies exist in several countries. Usually, regulatory bodies are founded as soon as more than one telecom operator emerges, but, even in a monopoly environment, a Regulatory Body can exist and perhaps control the offering of telecommunications services by the monopoly, handle the control of tariffs, and quality of the telecommunications services.

The regulations that the Regulatory Body is allowed to establish are within boundaries set by the laws and associated legal instruments, defined here in the legal framework for competition. There is no standard arrangement to determine if a regulation should be made by the Regulatory Body or by laws. Regulations from Regulatory Bodies are, however, more flexible than laws, since laws usually have to be made in a formal process involving, for example, a majority vote by the representative of the population in a congress or senate, whereas Regulatory Body regulations usually do not have to pass these barriers. The laws of a country, however, usually must provide a legal basis for the Regulatory Body to issue regulation in the first place [Melody 1997].

3.2.1 PURPOSE OF A REGULATORY BODY

In practice, we observe that Regulatory Bodies exist to promote "good competition" in the telecommunications services market, so that the consumers in that country can benefit from low prices and more choice of telecommunications services. Therefore, we describe "good competition" as competition where unequal advantages of competition are neutralized.

Two Examples of the Role of the Regulatory Body

The Regulatory Body can play an important role in the regulatory environment and can make or break the effects of the legal framework for competition. For instance, in the United Kingdom (UK) during the period between 1984 and 1992), the Regulatory Body in the UK (OFTEL) ruled that Cable and Wireless had to pay relatively high access charges, also called "interconnection charges" to British Telecom, for access to the local telephone lines, which were mostly owned by BT. The access charges were higher than British Telecom's cost were only on a per-minute-use basis and not on a per-call basis [Cave 1997]. The given cost structure made it impossible for Cable and Wireless to compete effectively. When this was changed by the Regulatory Body in 1992, Mercury gained a better cost basis and could begin to capture significant market share.

In another example; the Regulatory Body in New Zealand (the Ministry of Commerce), established very few regulations and had no enforcement power after the market opening in 1987, which resulted in years of battles between the dominant telecommunications operator Telecommunications Corporation of New Zealand (TCNZ) and new telecom operators on interconnection between networks. The first new telecommunications operator, Clear Communications, was connected with TCNZ more than three years after the first negotiations for interconnection started [Yankee 1997]. Technically, such a process would only require three to six months, although the bottleneck in these processes is usually political, rather than technical. In a later stage, the Regulatory Body was assigned more enforcement power and established more regulations to speed up negotiations for interconnection and access to TCNZs other infrastructures.

Regulatory Bodies in Various Countries

Some examples of regulatory bodies and some of their regulation in various countries are shown below [Oliver 1996, Clifford Chance 1997, Melody 1997, Noam 1997].

- *AUSTEL* Formed in 1989, the regulatory body of Australia regulates the telecommunications environment. On July 1, 1997, the PSTN telecommunications market in Australia was fully opened and went from a duopoly state to a competition state.
- *Ofkom* The regulatory body of Switzerland, was founded in 1997 and developed regulation for interconnection of PSTN services, which opened for competition January 1, 1998.
- *CRTC* (Canadian Radio-Television and Telecommunications Commission), formed in 1996, chose a liberalization of national long-distance communications first. International traffic opened for competition later, in October 1998. There are special rules for traffic to and from the U.S.
- *FCC* (Federal Communication Commission) FCC is the regulatory body of the United States and plays a key role in supervising competition in the U.S. For instance, it has the responsibility for implementing competition in local PSTN services, as described in the Telecom Act of 1996. Competition in long-distance PSTN services was introduced in 1984 and in local PSTN service in 1996.

- *MF PTE* (Ministère Français de Poste, Télécommunications et Espace) In France, the Ministry performs the regulatory body tasks until a regulatory body is founded. Competition in PSTN services was introduced on January 1, 1998.
- OFTEL (Office of Telecommunications) The regulatory body of the United Kingdom, was founded in 1982. OFTEL is largely independent of the government. It opened the United Kingdom for one new telecommunications operator in voice services (PSTN) in 1982, resulting in a duopoly (legal framework for competition score D) and for competition by an unlimited number of suppliers in 1994.
- *OPTA* (Onafhankelijke Post en Telecommunicatie Autoriteit) The regulatory body of the Netherlands, founded in 1997, is charged with implementing the higher-level directions given by the office of the Ministry or HDTP (Hoofd Directie Telecommunicatie en Post, General Directorate on Telecommunications and Postal services). It opened the market for voice (PSTN) services in July 1997 [Tempelman 1997].

As yet, there is no regulatory body for the European Union as such, but a body may be formed in the future. So far, the Directorate General IV and XIII of the European Commission have served as the most important sources for regulation. In the U.S., the FCC is a regulatory body that enacts regulation for all states, but individual states often also have regulatory bodies, usually called Public Utility Commissions (PUCs).

3.2.2 CATEGORIES OF THE REGULATORY BODY

Six categories of regulations have been identified for examining and quantifying the influence of the regulatory body in a country. Ideas for these categories came from practice* and various literature, such as the description of the Meta Telecom Maturity model [Johnson 1997]. The categories were chosen to be as much as possible independent of each other, but not all categories are expected to be completely independent, which should be taken into account when the scores are used in an analysis. The starting values for the scores are mentioned in the categories, and each of the categories is scored to carry equal weight. The scores for each category will be added and result in a total score for the Regulatory Body. The categories are numbered 1 to 6.

---

* AT&T Law and Government Affairs Department, spring 1998.

1. **Regulations concerning regulatory body independence and enforcement power**—This is a category that is not made *by* the regulatory body, but made *for* the regulatory body. Ideally, in order to be effective,* a regulatory body is considered a body that is independent of all telecommunications operators, with the responsibility of enacting telecommunications regulations that stimulate competition and quality of service. The regulatory body should have the power to enforce the rules that it establishes. Starting values were chosen for the following sample answers to questions a and b:

    a) Who is the regulatory body? One of the following answers is possible:

    ♦ The (dominant) telecommunications operator is the regulatory body: score 0 points.

    ♦ The (dominant) telecommunications operator is privatized (independent of government) and the regulatory body is a government department: score 1 point.

    ♦ The (dominant) telecommunications operator and the regulatory body are both independent of government: score 2 points.

    b) Does the regulatory body have *enforcement power*? One of the following answers is possible:

    ♦ Penalties or sanctions can be imposed for violation of regulatory body regulations**: score 1 point.

    ♦ Penalties or sanctions cannot be imposed for violation of regulatory body regulations: score 0 points.

For each of the two questions above, the scores are added to form the score of the category. This means that category 1, just like the other categories of the Regulatory Body, may have scores ranging from 0 to 3 points.

2. **Regulations concerning the licensing process**—If there is a need for parties to obtain a "license" before they are allowed to offer telecommunications services in a country, the regulatory body can influence competition by influencing the process that regulates the awarding of telecommunications licenses to parties. The regulatory body can determine that the licensing process fulfills the following criteria:

---

* Most regulatory bodies we encountered so far are government organizations.
** It is noted that procedures can be lengthy before penalties can actually be imposed and they can therefore be of limited effectiveness.

◆ The licensing process is reasonably *short*. As a guideline we establish that the average length of the total licensing process is less than six months. This is the total licensing process including a public appeal process. This criterion gets 1 point if applicable and 0 points if not applicable.

◆ The licensing process is *open* to the public in its criteria and process (transparent). This is reflected in requirements of openness for hearing sessions or publication in newspapers. This criterion gets 1 point if applicable and 0 points if not applicable.

◆ The licensing process is nondiscriminatory. The regulatory body treats all applicants in the same way. This criterion gets 1 point if applicable and 0 points if not applicable.

For each of the three criteria, the starting values chosen are 1 point and scores are added to get a score for this category ranging from 0 to 3. This means that, if the licensing process is reasonably short, open, and nondiscriminatory, the category scores 3 points.

The literature [Cave 1997] suggests that active use of the licensing process by the Regulatory Body can influence competition considerably, as in timing of competition on a per-service basis. This can be done by making the possibility to offer services dependent on other parties in the industry. The U.S. Telecommunications Act of 1996 allows local telecommunications operators (Regional Bell Operating Companies, or RBOCs, in the U.S.) to offer long-distance voice services, but only after the geographical market area of the RBOC has already encountered competition of new suppliers.

3. **Regulations concerning equal access to infrastructure**—With these regulations, the dominant telecommunications operator gives equal access to some of its facilities, including switches, cables, ducts, and antennas to telecommunications operators, including to the service providing the organization within its own company.* Equal access can be provided in various areas. We use three criteria that are each scored with starting values of 1 point if they are applicable and 0 points if not applicable:

---

* To carry out this regulation, often the dominant operator is split in terms of legal structure and accounting measures in a "service provider" and an "infrastructure operator." This way "cross subsidization" can be avoided between the "infrastructure operator," which operates in a monopoly environment as the only provider of infrastructure and the "service provider," which operates in a competitive environment.

a) Right to build a link/node infrastructure oneself. This includes the right-of-way, which means the right of a party (e.g., a new telecommunications operator), to lay cable in areas of land that the party does not own. Score is 1 point if applicable; 0 points if not applicable.

b) Access to link/node infrastructure (equipment) of the (dominant)* telecommunications operator is mandatory and based on internationally accepted technical standard interfaces. Score is 1 point if applicable; 0 points if not applicable.

c) The dominant infrastructure service provider is obliged to accept, e.g., co-location of equipment at its sites. Also, unbundling of services is applicable, such as unbundling of the local loop to the customer, which can be accessed by all operators. Score is 1 point if applicable; 0 points if not applicable.

Under a), the criteria are used similarly for all different kinds of infrastructure, such as equipment at links or nodes and cables, but also the use of common media, such as frequency allocations for wireless and satellite applications. For fast provision of services, often wireless links are used in both networks for "mobile operators" as well as "fixed infrastructure" operators. Therefore, access to frequency spectrum can be important, even when there is only competition allowed in wireline networks.** Adding the starting values of the three criteria in this category results in scoring for this category to range from 0–3 points.

4. **Regulations concerning price of interconnection**—The network for infrastructure services of the dominant telecommunications operator, or the entity that owns the infrastructure that provides the direct access to the locations of the subscribers in

---

* It is necessary to have access to an operator of facilities that are a "bottleneck," meaning facilities that belong to that particular operator AND that are needed to realize the intended provisioning of the service.

** "Wireless" and "wireline" (or "fixed") services of operators are usually defined on the basis of what the end-user sees. The end-user with a wireless handset that communicates directly with the operator (using, e.g., GSM) considers himself on a wireless service (e.g., by a "wireless operator"). The end-user with a wireline handset considers himself connected to "wireline" service of a "wireline operator." The actual transmission of the signal from user A to user B can use various media to complete its path, e.g., including satellite, terrestrial wireless, fiber cable or copper cable. Access to all infrastructure (to both wireless frequency spectrums, as well as equipment and cables of the fixed network) is essential for both kinds of operators to establish service.

the country, should be open for connection (also called "inter-connection") with other telecommunications operators. The regulations concerning interconnection are identified in this category by the conditions a, b, and c.

a) The price and conditions for access to infrastructure are cost-justified. Starting value: 1 point if applicable or 0 points if not applicable.

b) The price and conditions for access to infrastructure are known to the public. Starting value: 1 point if applicable or 0 points if not applicable.

c) The price and conditions for access to infrastructure are exactly* the same for all competing telecommunications operators, including the competing part of the dominant telecommunications operator that uses the network infrastructure. This criterion is referred to as "nondiscriminatory." Starting value: 1 point if applicable or 0 points if not applicable.

For each of the three regulations, the starting values chosen are 1 point and scores are added for the category, leading to a3-point score if all regulations are present or less if some are not present. Total range of scores is 0–3 points.

The importance of these interconnection regulations is shown in various practical cases (see the BT/Cable and Wireless example shown earlier in this chapter). Cost-justification (condition a) is an extensive subject that requires detailed accounting rules. In particular, the allocation of fixed costs of a network (which are very high in a capital-intensive type of operation) is a detailed accounting process and is often challenged in court cases between existing telecommunications operators and new telecommunications operators. Most literature on this subject is available in jurisprudence [Oliver 1997].

5. **Regulations concerning "fair competition"**—There are measures that can promote the fairness of competition by establishing regulations aimed at particular parties. Depending on the nature of the competition, some of these measures can be used temporarily or permanently. The two kinds of regulations addressed here are: price regulations, listed under a) and universal service, listed under b). Opinions vary widely on the effectiveness of price regulations

---

* Exactly the same may be difficult in reality, e.g., as the availability of co-location space cannot be guaranteed for all players that like to position their equipment next to that of the dominant operator.

[Cole1991, van Cuilenburg 1995]. We have chosen starting values, which increase when fewer regulations are imposed.

Universal service is a regulatory criterion that has been estimated in literature to promote fair competition, as it equals the obligations for telecommunications operators that are offering services [Hammond 1998]. The two kinds of regulations concerning "fair competition":

a) Price regulations.

Any of the following three situations may exist:

♦ Prices for services are established with government approval: score is 0 points.

♦ Price regulations applicable (for instance minimum or maximum prices to be charged for certain services): score is 1 point.

♦ No price regulations: score is 2 points.

b) Universal service. Universal service means that telecommunications operators have to offer their services in all areas of a country at the same price and under the same conditions or must pay into a "universal service fund" to subsidize the provision of service to certain nonprofitable customers to which they don't provide service. Universal service regulation is made because telecommunications operators would otherwise not offer service to low-revenue customers or customers in areas where there is a high cost of provisioning the service. Starting values are as follows: 0 points if universal service does not exist, 1 point if universal service exists in some form in a country.

A description of universal service in more formal wording is: *Universal service is the offering of an afforable service on a nondiscriminatory basis, with the same quality and price, independent of where the user is requesting the service.* Today, universal service exists in most countries for basic telephone service (public switched telephone network), but only in some places for other telecommunications services. As universal service often plays a role in lawsuits between telecommunications operators, we provide a more detailed example of how universal service can be implemented.

Example

An implementation of universal service can be found in the U.S. Universal service in the U.S. is regulated in various laws, including antidiscrimination laws, as well as the Telecom Act of 1996, section 254 [US government 1996], which provides for special funds to be financed by all telecom operators for the purpose of maintaining universal service for various types of services. Also, there are special programs with government subsidies that can give access to basic services. Increasingly, access to PSTN (in the law described as "basic telephony") is no longer considered sufficient in U.S. politics. Access to advanced telecommunications services such as the Internet is seen as another service that all people must be able to have [Oliver 1997]. As regulations for universal service for the access to basic telephony, several programs, such as the "Lifeline" and "Link-up" programs, exist to actually subsidize the users of telecommunications services. For access to advanced telecommunications services, a multitude of regulations exist, such as antidiscrimination laws, as well as specific programs to assure universal service for certain groups in the population, such as members of schools and libraries. Table 3.3 gives an overview of current universal service regulations in the U.S. There are regulations in several states that supplement these.

TABLE 3.3
Example: Universal Service Regulations in the U.S.

Subject:	Applicable Regulation
Access to Basic Telephony	"Lifeline" and "Link-up" programs (subsidies for low-income households) "Antidiscrimination laws"
Access to Advanced Telecommunications services	Regulation based on sec. 254 and 706 of the Telecom Act of 1996, "Universal Service in all regions" Regulations that arrange targeted subsidies for: Schools and libraries Rural Health care providers Community Based Organizations

The scores of category 5 will be tracked both as an aggregate sum to represent the whole category, and individually to be able to do more-detailed statistical analysis in Chapter 9. With the starting values chosen, the scores in the category range from 0–3.

6. **Regulations concerning number portability**—This category handles regulations on number portability or subscriber identification and numbering. An important aspect of subscriber iden-

tification and numbering is number portability, the ability to keep one's current phone number, domain name, or TCP/IP address while switching telecommunications operators or service providers.

Number portability exists in different forms. We will assess the "local number portability," which concerns the possibility for subscribers to keep their own numbers or identification. Number portability has been shown to be an important criterion for subscribers and very few users want to give up their numbers or identification to get a lower price or better service from a different telecommunications operator [Franx 1998, Terplan 1999]. Number portability scores 3 points if it is implemented and 0 points if not. The number portability requirement is fulfilled when the numbering plan is administered by an independent body or department, to assure fair treatment of new telecommunications operators that request identification numbers.

3.2.3 TOTAL OVERVIEW OF THE SCORING FOR THE REGULATORY BODY

For each of the six categories of regulatory body, we will use the scores from 0–3, as mentioned in the categories. The starting values were chosen to cover all categories about equally and in such a way that a higher score is expected (according to literature [VanCuilenburg 1995] and personal experience), which should result in more, lower-priced, or faster-provisioned services. The weighting may be changed after the statistical analysis that we plan to do in case studies. The answers are scored and added per country in the right-hand column in Table 3.4 The total score is the addition of the category scores and is expected to give an indication of the total regulations of the regulatory body, which means that the higher the score, the more the regulatory body is expected to promote "good competition" as referred to earlier in this section. Each of the categories has a score ranging from 0–3, so the total score for the regulatory body in question should range between 0 and 18.

To describe the regulatory bodies of multiple countries, the format of Table 3.4 can be confining, as only a few columns fit across a page. Therefore, the figure can also be depicted in a "condensed horizontal format," as shown in Table 3.5. The condensed horizontal format does not allow room for the scores of each of the individual bullet items (criteria) within the categories. An exception is made for category number 5, fair competition regulations,

TABLE 3.4
Regulatory Body Scores for Sample Country X

Regulatory Body	Country X	
	Status (starting value)	Category (starting values)
Categories (with the point scores in brackets)		
1. Independence/Enforcement power		
1a. Regulatory body independence. One of the following:		
=> dominant telecommunications operator is regulatory body (0 points)		
=> regulatory body is government department and telecommunications operator privatized (1 point)	Independent	
= >independent telecommunications operator and regulatory body (2 points)	(2)	
1b. Does regulatory body have enforcement power? (Y=1, N=0)	Y (1)	3
2. Licensing process		
Licensing process is:		
• short (Y=1, N=0).	N (0)	
• transparent (Y=1, N=0).	Y (1)	
• nondiscriminatory (Y=1, N=0).	N (0)	1
3. Equal access to network infrastructure		
A new telecommunications operator is allowed to:		
• build network infrastructure (Y=1, N=0).	Y (1)	
• have access to network of a dominant (bottleneck) operator (Y=1, N=0).	Y (1)	
• require mandatory co-location from a dominant operator (Y=1, N=0).	Y (1)	3
4. Price of interconnection		
Access charges are:		
• cost-justified (Y=1, N=0)	Y (1)	
• published (Y=1, N=0)	Y (1)	
• nondiscriminatory (Y=1, N=0)	Y (1)	3

continued

TABLE 3.4 (CONTINUED)
Regulatory Body Scores for Sample Country X

Regulatory Body	Country X	
Categories (with the point scores in brackets)	Status (starting value)	Category (Total)
5. Fair competition		
(a) Price regulations set the following restrictions:		
= Services priced only with government approval (0 points)		
= Price caps (minimum/maximum prices) (1 point)		
= No price regulations (2 points)	no reg(2)	
(b) Universal Service or fund contribution is mandatory (Y=1, N=0).	N (0)	2
6. Number portability		
Number portability mandatory (Y=3, N=0)	Y (3)	3
TOTAL Regulatory Body Score		15

Source: Van den Broek, F., adapted from an earlier figure published in *International Journal of Network Management*, vol. 7, John Wiley and Sons, 1997. With permission.

since that category is composed of two very different criteria—price regulations and universal service. Splitting the scores of category 5 makes it easier to analyze potential relationships among these criteria when applying the model in step 5 of the research methodology.

3.3 COMPETITION ACTIVE

The third element of the regulatory environment is "competition active," or the number of competitors present for a telecommunications service in a country. This element differs from the other two in that it is *descriptive* rather than *prescriptive*. It is part of the regulatory environment merely to confirm that the two other elements actually do (or do not) result in competition. The presence of many (legally operating) competitors in a country shows that the regulatory environment has created "low" barriers for entry, a sign of a competitive market. The number of competitors offering a service at a level in the service value model are counted, forming

TABLE 3.5
Regulatory Body Scores Shown in Condensed Horizontal Format

	Regulatory Body							
	1. Independence/ Enforcement power	2. Licensing process	3. Equal Access to Infrastructure	4. Price of interconnection	5. Fair competition regulations		6. Number portability	TOTAL Regulatory Body score
					Price Regulations	Universal service		
Country X	3	1	3	3	2	0	3	15
Country Y	2	3	2	3	1	1	0	
Country Z	2	3	2	2	0	0	0	9

the basis for the score. A competitor is only counted as such if its market share is greater than a threshhold, which is determined by what the fragmentation of the market is. Formulas for "information entropy" are used [Theil 1997] and general threshholds are between 1% and 5%. The market share number turns out to be essential, as it often happens that many competitors start in a certain market, but, in some cases, competition is so tough that the new entrants get almost no market share. The starting values for the scores are chosen so that the number of points is in line with the experience of expected competitiveness. "More than two competitors" denotes a competitive market and is therefore awarded 3 points, with lower scores for fewer competitors. This results in the scoring as shown in Table 3.6.

The competitors are counted for each of the three layers in the service value level and the scores for all of the three layers in the service value model are then added by country. Table 3.7 shows an example for the scoring of the element "competition active."

3.4 OVERVIEW OF THE REGULATORY ENVIRONMENT

When the three elements described as part of the regulatory environment are taken together, a regulatory environment can be depicted as a combination of

TABLE 3.6
Starting Values for Competition Active

Competition active	Code	Starting value
No competitor	0	0 points
One competitor	1	1 point
Two competitors	2	2 points
More than two competitors	>2	3 points

TABLE 3.7
Scores for the Element "Competition Active"
for Countries X, Y, and Z

	Services						Total competition active
	Value Added		Basic Data Transfer		Infrastructure		
	Number	Score	Number	Score	Number	Score	
Country X	> 2	3	1	2	> 2	3	8
Country Y	1	2	1	2	> 2	3	7
Country Z	> 2	3	> 2	3	1	2	8

the tables shown in the previous section, and is shown with scores for sample countries X, Y, and Z in Figure 3.2.

In the regulatory environment model, each of the three elements fulfills a "role" that is probably best explained by viewing each of the three in a stage of competition active in a particular country. When the country opens markets to certain services for competition, laws are often first made to allow this. Later, more-detailed regulations are made (usually by a regulatory body), and finally, competitors eager to take advantage of the newly created opportunities will begin to appear.

The information needed to fill in the boxes in the regulatory environment model can be obtained from various sources. Examples are geographical area studies, such as those done by Noam [1992, 1994, 1997], high-level overviews that can be found in handbooks [Frieden 1996, Terplan 1999], or consultant reports and reports of official organizations [OECD 97] The scores

	Legal Framework for competition						
	Value Added Services		Basic Data Transfer Services		Infrastructure Services		TOTAL Legal Framework
	Code	Score	Code	Score	Code	Score	Total score
Country X	C	3	M	0	M	0	3
Country Y	C	3	C	3	C	3	9
Country Z	C	3	C	3	C	3	9

	Regulatory Body							
	1. Independence / enforcement power	2. Openness and length of licencing process	3. Equal Access to Infrastructure	4. Price of interconnection	5. Fair competition regulations		6. Number portability	TOTAL Regulatory Body score
					Price Regulations	Universal service		
Country X	3	1	3	3	2	0	3	15
Country Y	2	3	2	3	1	1	0	12
Country Z	2	3	2	2	0	0	0	9

	Competition active						
	Value Added Services		Basic Data Transfer Services		Infrastructure Services		TOTAL
	Number	Score	Number	Score	Number	Score	
Country X	> 2	3	1	2	> 2	3	8
Country Y	1	2	1	2	> 2	3	7
Country Z	> 2	3	> 2	3	1	2	8

FIGURE 3.2 Example of scores in the regulatory environment model.

are added up on a per-country basis to arrive at a total score of the regulatory environment. In the example in Figure 3.2, Country X would be rated according to the sum of all the Country X columns—3 + 8 + 15 = 26 points. The resulting total score of the regulatory environment per country can also be used as an input parameter for the cost-effective management model, which will be developed in Chapter 5.

3.5 NEW DEVELOPMENTS IN REGULATORY ENVIRONMENTS

In February 1997, several member countries of the World Trade Organization (WTO) signed an agreement called General Agreements on Trade in Services Concerning Basic Telecommunications, often referred to as the WTO Basic Telecom Services Agreement [Oliver 1998] to commit each other to opening their telecommunications markets to competition.

Opening the markets for competition would happen according to a time schedule that was proposed by each of the countries that signed the agreement. Each had an "offer," which consisted of a series of plans, including a date to open the market, categorized mostly by a service value level or a similar description. The four main categories of services that are covered by the WTO Basic Telecom Services agreement are domestic public switched telephone network (PSTN), domestic long-distance PSTN, international PSTN and satellite services. The countries that signed the WTO Basic Telecom Services agreement represent 79 percent of the world economy*

TABLE 3.8
Example of "Offers" of a Few Countries in the WTO Basic Telecom Services Agreement

Situation regarding a service	Greece	Spain	Venezuela
Competition in domestic PSTN	in year 2003	in 12/1998	in 11/2000
Competition in international PSTN	in 12/1998	in 12/1998	in 11/2000
Possibility to own infrastructure	in year 2003	in 12/1998	in 11/2000
Allows bypass into country	in year 2003	in 1/1998	not planned

Table 3.8 shows an example of "offers" of a few countries that were involved in the WTO Basic Telecom Agreement.

As shown, most offers contain commitments regarding competition for particular services. In their offers, countries state a date by which they promise competition will be allowed for that particular service. The WTO Basic Telecom Services agreement itself then states requirements for when opening each of the services must actually happen. For example, when a country declares that competition exists in PSTN services, the interconnection of competitive telecommunications operators to the PSTN network must be nondiscriminatory and cost-based. There are numerous developments in the regulatory environment of many countries, and individual country organizations as well as supranational organizations such as WTO and EU are developing regulations to standardize the markets.

3.6 SUMMARY

The regulatory environment has been described by modeling it in a way that determines three elements: legal framework for competition, regulatory body,

---

* This is in terms of GNP; however, in terms of world population, they represent only 19% of the population.

and competition active. Each of the elements assesses the situation in a country and defined starting values are assigned to express these situations in a quantitative set of scores. The combination of the scores creates the "regulatory environment model." Thus described, the regulatory environment of a country can be expressed in a quantitative way so it can be used as input for the cost-effective management model, together with the telecommunications services offering, which will be described in Chapter 4.

4 Telecommunications Services Offering

The telecommunications services offering, which is discussed in this chapter, is the second input parameter for the cost-effective management model under study in step 4 of the research methodology. An international network might use and/or deliver telecommunications services, which are part of the telecommunications services offering in a particular country. In this chapter, services will be categorized in different ways in section 4.1, and aspects of telecommunications that are valued by international organizations are studied in section 4.2. Services that are important for international organizations are outlined in section 4.3, and a telecommunications services offering matrix is developed in section 4.4 to model the offering in a particular country.

4.1 TELECOMMUNICATIONS SERVICES

History of Telecommunications Services

Communication across a distance of more than a few miles was made possible by the invention of electricity-based telecommunications services by Samuel Morse in 1838. He invented the telegraph as a device that used electrical signals transported over a copper wire, and developed the Morse alphabet as a standard way of expressing language in electric signals. Later, the telegraph was also fitted to use radio waves for transmission.

The telephone, invented by Alexander Graham Bell in 1879, made it possible to communicate by voice. The basic operating principle of the telephone has changed little between then and now, but the technology for transmitting the electromagnetic signals from the telephone has changed drastically. Initial communication was based on overhead wires that all connected to a manually operated exchange, introduced in most countries around 1900. The first electromechanical switches appeared in the 1930s. They allowed subscribers to dial their own numbers, thus replacing the need for telephone switch operators. However, in most Western countries, it wasn't until the 1960s that all the manually operated switches were replaced by automatic switches—in fact, some countries still rely on manually switched networks today. Since 1966, when the first telecommunications satellite was launched, telecommunications between continents have grown enormously and international transmission of television images has become possible.

Introduction of Multiple Suppliers

Historically, creating telecommunications services in individual countries was a task of each government. The governments operated a ministry or public body that was responsible for offering telecommunications services, which were seen as utilities that should be provided to everyone at the same price, usually regardless of the location where the user was located. There were also technical reasons to keep the complete network in the hands of one party. The technical solutions had not been found to enable the use of a "bottleneck resource," such as the infrastructure, by more than one party.

The service offering consisted of a limited number of services that the government deemed necessary and beneficial for the users in the country. The government also controlled the quality of the services offered. If services were offered by a separate state owned company or one that was not state owned but considered monopolist, requirements for the telecommunications services were placed on the company. These one-supplier environments still created problems. In countries where that situation still exists, the examination of a telecommunications services offering in that country is an easy task, as only one information source is needed.

In the early 1980s, some governments started to change that situation and multiple suppliers emerged in a few countries from that time and in more countries into the 1990s.

Telecommunications Services

Telecommunications services can be categorized in different ways. Distinctions that are often used are public versus private services, data versus voice services and wireline versus wireless services. Most, but not all, services can be categorized with these distinctions.

Figure 4.1 shows an overview of private networks, and public and private networks based on public networks, which fit none of the categories. Services of both public and private networks are commonly used by international organizations. The difference, in essence, stems from the ownership of and access to the network, which can be public (a group owns it and everyone has access to the same network) and private (only

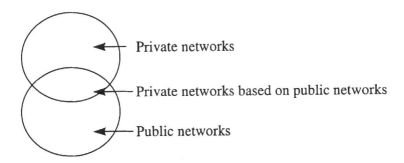

FIGURE 4.1 Private networks, public networks, and private networks based on public networks.

the party that leases or buys the network can access it). In practice, there are also services that are private, but are implemented on public facilities. They are called "virtual private networks." In that case, the telecommunications operator usually implements the technological measures on a public network to give the virtual private network user access to a seemingly private network.

Service Value Levels and Supplier–Customer Relationships

Telecommunications services can be categorized in service value levels as described in section 2.3. The main purpose for categorizing services in such a way is that many countries use a similar distinction to classify services.

In reality, there might be multiple parties (and supplier–customer relationships) at each layer of the service value model. Figure 4.2 shows an example of actual supplier–customer relationships, described in terms of the service value model in a two-country situation. Infrastructure operators at the bottom of the service value model actually own and manage the border crossing links. Some value added services providers try to become "international' by offering "managed services" at, for example, the value added services level, such as the value added service provider (d), which is operating in both country A and country B and therefore maintaining relationships with operators (e) and (f), both of which obtain their services from infrastructure operators (g) and (h). These operators do not have an individual network that crosses international borders. They do have "bilateral relationships" with each other. In such a case, the international link is usually jointly owned by both infrastructure operators, with each responsible for management of half the link—the part that is within their country. The service value model depicted helps to create an understanding of the relationships among the parties that enable the operation of the international

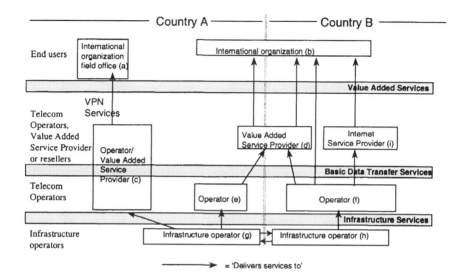

FIGURE 4.2 Supplier–customer relationships described in terms of a service value model.

network. In Figure 4.2 this is shown where both the end user (the multinational organization (b)) and the value added service provider operate across international borders. (b) also uses Internet services from Internet service provider (i), which is obtaining TCP/IP services from operator (f), which in turn is connected with other operators using infrastructure operator (h).

Standardization

The *telecommunications services offering* can be categorized by the types of services, which can include, for example, video, voice, and data services. Some services are standardized and named, so that a wide audience can understand the service and use it with standardized equipment and peripherals that connect well with the networks [Tanenbaum 1997, Looijen 1998]. Examples of standard services are: X.25; ETSI EURO ISDN Q931, a packet switched service; and a digital circuit switched service according to the standards of ETSI (European Telecommunications Standards Institute).

4.2   ECONOMICS OF SERVICES CROSSING INTERNATIONAL BORDERS

An important difference between the management of a network *with* or *without* international links, is the price for telecommunications services that the international organization must pay*. Two main observations were made in studying both situations:

1. An international link is much more expensive than a national link.
2. The price of links varies very much depending on the country in question.

Regarding the first observation, currently everywhere in the world**, the prices for services that cross an international border are much higher than for the same service that does not cross such a border. In fact, an international line of a certain distance is usually 3–5 times*** [Molony 1999] more expensive than a similar leased line of the same distance that does not cross an international border. From an economic point of view, this could render certain applications impossible for the end-user to implement internationally, whereas those applications could have been cost-effective when not required to cross international borders, such as in large countries like the U.S.

A graphical depiction of this multiple for a set of countries is shown in Figure 4.3 [Molony 1999]. For each country, the left bar shows the multiple of international versus national leased lines in the year 1997 and the right bar shows the multiple for 1998. In most cases, the multiple remains between 2 and 5, and does not decrease, but rather increases, except for Belgium and Denmark. This shows that international networks seem to become more and more "different" from noninternational networks, at least in terms of costs of telecommunications services.

---

* Leasing is the most common way of acquiring capacity for international links. If the regulatory environment allows it, there is a possibility to buy (parts of) links of infrastructure that crosses international borders. This can be done in the form of IRUs (indefeasible rights of usage) or by actually buying (a share of) a cable facility or satellite. It may be possible that buying instead of leasing decreases the difference in costs between international and national links. Since buying is possible for only a very select few companies (because of regulatory and capital constraints), it is not considered an alternative.
** Some improvements can be expected from international alliances of telecommunications operators. So far, however, they have not been able to force major changes in international tariffs.
*** Cost comparison by BTG, The Telecom User group, The Netherlands, 1999. For example, a 2Mb/s leased line that crosses a border between The Netherlands and Belgium is 2.5 times the cost of a leased line with the same capacity and the same distance within The Netherlands. International leased lines pricing overview of Eurodata Foundation.

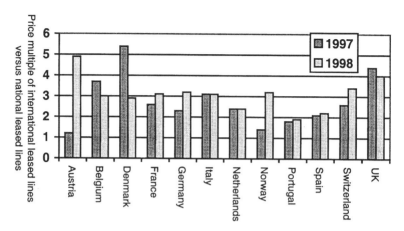

FIGURE 4.3 Price difference (multiple) of international versus national leased lines in 1997 and 1998 for selected countries.
Source: Molony [1999], INTUG

Regarding the second observation, there are big differences in price in different countries. An example is shown in Figure 4.4, which shows the market prices of leased lines within various countries.

The data in Figure 4.4 is a comparison of tariffs within a country [Eurodata1998]. The most expensive country for leased lines (Italy) is 15 times more expensive than the cheapest country (Sweden).

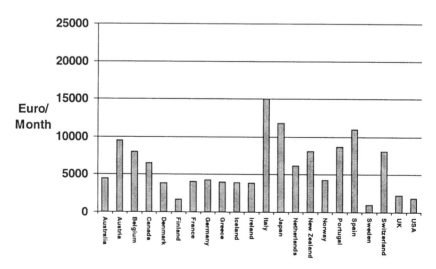

FIGURE 4.4 Monthly costs of 100km, 2Mb/s leased lines in several countries.
Source: Eurodata [1998])

Infrastructure Operators' Pricing

There are many reasons for the differences in prices that infrastructure operators demand for international leased lines and other international services compared with similar services that do not cross borders. Many infrastructure operators worldwide still do not pass on the benefits of lower-cost technology to the consumer of the service, but decrease their prices at a much slower pace. Even though there may be multiple suppliers of services at the VAS-level or BDTS-level, infrastructure services are still a resource offered by only one party in many countries. The owners of the infrastructure are therefore often still in a position to charge higher prices. Infrastructure services crossing international borders is a scarce commodity in many countries, as laying cable requires a *right of way* (the right to dig and lay a cable in a piece of land) in separate countries up to the international border. The actual land close to the border is often owned by the state and rights of way can be difficult to obtain. Transmission by radio waves across an international border is also possible, but requires licensing for use of radio spectrum in countries on both sides of the border.

International Settlements

Another background that might explain the high prices is that the international arrangements between infrastructure operators in different countries still include high reimbursements, also called "settlements," between different countries for the completion of each other's telecommunications services in the terminating country. These international settlements could, in theory, be close to zero, if there are as many leased lines or as much traffic from country A to country B as from country B to country A. In that case, it doesn't matter how high the per-minute or per-capacity charge for cross border traffic is. In reality, however, the balance of originating and terminating traffic between most pairs of countries in the world is tilted toward one side. The country with the lowest prices of services to the users often generates more originating traffic or requests for capacity and thus ends up paying the terminating country for the difference between the amount of traffic sent and the amount received [Frieden 1995, Sandbach 1996]. Graphically settlement payments are shown in Figure 4.5. The leftmost bar shows 50 million minutes of traffic sent per year from country A to country B and the rightmost bar shows 30 million minutes of traffic incoming from country B to country A. The difference, also called "unbalanced traffic," triggers a settlement payment for 20 million minutes from

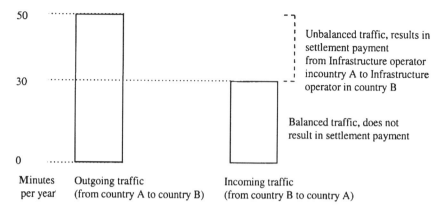

FIGURE 4.5 Unbalanced traffic and settlement payments between countries A and B.

infrastructure operator(s) in country A to infrastructure operator(s) in country B. (Note that this is an example and the numbers of minutes have been chosen arbitrarily.)

Nontraditional International Traffic

"Non-traditional international traffic" is a euphemism often used in telecommunications for new ways for operators to avoid paying the high international settlement cost for voice traffic. Important methods are to "bypass into a domestic network" and "refile" or "reorigination." Usually, this practice is used by infrastructure operators, but could even be done by anyone with the right telecommunications equipment.

Bypass into a domestic network—uses an international leased line (or another international means of transport, such as the Internet) to bring traffic from a foreign country into the termination country and end it in the domestic network at local rates. Termination can be done through organizations with high capacity "links" into the local network, such as hotels or large organizations with relatively large volumes of traffic and private exchanges. The traffic can then be disguised to look as if it comes from the hotel, but, in reality, it is foreign-originated traffic that avoids the international settlement payments. A difficulty with bypass is that quality is often poorer than if traffic had been routed through an official telecommunications operator. Also, this method can be easily detected when the telecommunications operator in-country realizes that the hotel suddenly sends a lot more traffic than it usually does. Other methods of bypass are, for instance, based on voice-over-Internet technology.

Refile or reorigination—entails sending international traffic, in particular, voice traffic, to the destination country via a detour, a third country, which cooperates by labeling the traffic as if it had originated in the third country and then forwarding it to the termination country. The quality is usually higher than with bypass, but depending on the length of the "detour" and the communications technology (satellite, microwave, or fiber cable), this method can introduce propagation delays. Reorigination is the same activity as refile, but viewed from the third party. So, for example, party A refiles traffic to party B via party C, which is reoriginating this traffic to party B. More details on non-traditional traffic are available in the literature [Scheele 1998].

4.3 INTERNATIONAL ORGANIZATIONS AND TELECOMMUNICATIONS SERVICES

International organizations, which use telecommunications services at a rate growing by 15% per annum, spent more than $100 billion in 1997 on telecommunications services [AT&T 1997]. In this section, telecommunications services for international organizations are examined in more detail as a preparation for the creation of the matrix, which we will call the *telecommunications services offering matrix*. Our examination is based on the "network managers survey," which was conducted specifically for this research.

Network Managers Survey

The survey for this research project was carried out among the network managers of 17 international organizations in Europe*. It shows which services are most important to international organizations and what aspects of the services are most important. It furthermore focuses on the use of VPN services for purposes beyond the scope of this research [vanDalen 1995]. It showed that the following services are considered most important:

---

* Demand by multinational corporations survey among 17 international organizations, conducted by means of interviews and paper inquiries, AT&T/Lucent Technologies Netherlands, Edwin van Dalen. Organizations: Philips, DuPont, ING Netherlands, Shell, Nedlloyd lines, Netherlands postoffices, Hewlett-Packard, IBM, DSM, Monsanto, Akzo Nobel, Dutch Railways, Schiphol Telematics, PTT Post, AT&T, Ministry of Health and Environment, Unilever Meat Group. It should be noted that most of the respondents operate networks that cover many countries, but about half of the networks had more than 50% of the links concentrated in western Europe.

- Public Switched Telecommunications Network (PSTN)
- Integrated Services Digital Network (ISDN)
- Services many organizations use to establish a private network, such as:
    ♦ Leased lines 64kbps
    ♦ Leased lines 2 Mbps
    ♦ Packet Switched Data Network (PSDN). There are various standards, primarily X.25, but this may in the future grow to TCP/IP or asynchronous transfer mode (ATM)
    ♦ Virtual private network services, requested by the respondents primarily to combine some of the services mentioned above on one network and reduce costs of operators

The results of the survey also show that the most important aspect of these services for the network managers was price. Listed in order of decreasing importance are:

- Price of the services
- Geographical coverage area of the services
- Time, including: time of provisioning, Mean-Time-To-Repair (MTTR), Mean-Time-Between-Failures (MTBF), availability time of the network
- Reliability, indicated in particular for virtual private network services
- Security, indicated in particular for virtual private network services

Concerning this list of aspects, some remarks were made in the surveys by the network managers that were combined with our observations. For example, security was mentioned in particular concerning virtual private networks and wide area networks, but it is difficult to measure. An explanation given by one of the respondents was that "virtual private networks and ISDN networks are seen as a new technology and the fact that physical resources are shared with other users may introduce an initial fear for security problems." Reliability and security were not of high priority, but some network managers indicated that they just assume that both are provided adequately by the telecommunications operators. Other network managers simply indicate that they have adequate measures in place to guarantee reliability and security (such as redundant routing for reliability and encryption for security) and that they don't expect problems in those areas. An overview of the survey questions can be found in the Appendix.

Developing a Telecommunications Services Offering Matrix

The telecommunications services offering matrix that is to be developed should be easy to use and therefore include a limited set of aspects and services. The aspects that were mentioned by the network managers most often in the survey, area of availability and price, were taken and quantified to be used as input for a model that describes the telecommunications services offering in a country.

Area of availability

- Coverage, or the percentage of populated areas of the country where the service is offered. This aspect is detailed in section 4.4.1.

Price of telecommunications services, consisting of:

- One-time price of the service, such as installation fee
- Usage price of the service per unit of information or time

This aspect is detailed in section 4.4.2.

We will express the geographical coverage in a numeric value, based on the percentage of the geographical area. For price, we compare a standard set of services and usage, as defined in the OECD report [OECD 1997].

Other Aspects and their Representation

Apart from the two chosen aspects, a third aspect, "time," is not directly represented in the telecommunications services offering matrix. It plays a minor role and is only considered if the "time" aspect is much different from a normal situation. This is explained as follows: we will consider a service as simply not available if the "time" aspects of the service are so much worse than average that most international organizations cannot use them. Time aspects include downtime, provisioning time, MTBF, and MTTR. Downtime is the time the service is unavailable per month and provisioning time is the time it takes to provide a service to a new user. MTBF is the average time between two failures in a network. MTTR is the average time it takes to repair a failure.

For "downtime," we consider a service "unavailable in that geographic area" if it is available less than 90% of the time or if the provisioning of such a service takes more than four times the average provisioning time for those services in other countries. Area of availability and price of telecommunications services together form the telecommunications services offering matrix.

4.4   DEVELOPING A TELECOMMUNICATIONS SERVICES OFFERING MATRIX

The telecommunications services offering matrix is used to quantify the telecommunications services offering in the countries where the network operates. In order to fill the matrix we give the countries a "score," depending on the area of availability and the price of telecommunications services.

The main source for selecting the services that would be represented in the matrix were those used by the 17 multinational organizations that were surveyed. The following services, listed in section 4.3, were selected:

- International virtual private network service (VPN)
- International public switched telecom network (PSTN)—voice, fax, and modem service
- International integrated services digital network (ISDN)
- International leased line 64Kbps speed
- International leased line (digital) 2 Mbps speed
- International packet switched data network (PSDN)

An optional service to be selected could be international dark fiber service, which will only become part of the telecommunications services offering matrix when a sufficient number of countries offer it. Currently, the number of suppliers of this service is so small that it was not a candidate for the telecommunications services offering matrix. The service is only mentioned here because it represents the service value level "infrastructure service." The use of dark fiber services requires a license in many countries.

4.4.1   AREA OF AVAILABILITY

The geographical area in a country where the service is available can roughly be measured by determining in what cities and rural areas the service is offered in relation to the total size of the country. Apart from PSTN services, for which "universal service" regulations often exist, most services are not offered throughout the country. If service is available only in a particular (e.g., the capital) city, it can have strong repercussions on the cost of management of the network. This means that, for expansion into other cities, different—possibly more expensive or lower-quality—services would be required. An example is leased-line service, which may not exist in part of a country. Instead, an international organization then might be required to use very small aperture terminals (VSAT), which are used for setting up a leased line via satellite. In our matrix, starting values were chosen to represent three categories of area of availability as depicted in Table 4.1. A high score is expected to have a positive effect and thus lower the cost, as the services would be available in more areas, so more choice is available for choosing the type of service with the lowest cost.

TABLE 4.1
Scoring on Area of Availability

Area of Availability	Score
"+" = the service is available in almost all populated areas (80%–100% of the populated area)	3 points
"0" = the service is available in limited areas only (20%–80% of the populated area)	2 points
"-" = the service is available in very limited areas or is not available (<20% of the populated area or the service requires an unusually long provisioning time)	1 point

---

The starting values were chosen, so points range between 1 and 3. The "very limited or not available" category receives one point, as it represents very little or no availability.

4.4.2 PRICE OF TELECOMMUNICATIONS SERVICES

The price of the telecommunications services is measured relative to the prices of the same services offered in other countries in the world. For each of the six services that are part of the telecommunications services offering matrix, countries around the world are divided according to price into three groups.

1. Twenty percent of the total group represents the highest-priced countries for a certain service.
2. A second 20% represents the lowest priced countries for the same service.
3. The "middle group," or 60%, represents the remaining countries.

It is not always possible to use this division accurately, because it implies that all the countries in the world must be examined for their price of telecommunications services. In the test cases, the 25 OECD countries were taken as a reference group representing the world, which means that, for instance, the second group would mean the five lowest priced OECD countries for that service. Starting values were chosen as 1 point for the highest priced group, denoted with an "H," 2 points for the middle priced group, denoted with an "M' and 3 points for the lowest priced group, denoted with an "L." The scoring of price of telecommunications services is shown in Table 4.2.

TABLE 4.2
Scoring on Price of Telecommunications Services

Price of telecommunications services	Score
L = Low price (lowest priced 20 % of telecommunications operators on world market)	3 points
M = Medium price (average priced group of telecommunications operators on world market) The middle group of operators that are not in "H" or "L" categories.	2 points
H = High price (highest priced 20 % of telecommunications operators on world market)	1 point

---

As in area of availability scoring, the total score of price of telecommunications services ranges from 1–3.

In the current starting situation, all six services are counted as equally important, however, depending on the type of organizations that actually use the services, their importance may differ.

4.4.3 OVERVIEW OF THE TELECOMMUNICATIONS SERVICES OFFERING MATRIX

Using the scoring in the previous sections, Table 4.3 shows an overview of an example of the total telecommunications services offering matrix, which is the combination of the price of telecommunications services, using Table 4.2 as a reference, and the area of availability scores, using Table 4.1 as a reference. The scores of area of availability and price of telecom-

TABLE 4.3
Telecommunications Services Offering Matrix for One Country with Sample Scores

Type of Service	Country X				
	Area	**Score**	**Price**	**Score**	**Total**
International VPN service (VPN)	+	3	H	1	4
International Voice, Public Switched Telecom Network (PSTN)	0	2	H	1	3
International Integrated Services Digital Network (ISDN)	0	2	M	2	4
International Leased line (digital) 64Kbps speed	+	3	H	1	4
International Leased line (digital) 2 Mbps speed	+	3	M	2	5
International Packet Switched Data Network (PSDN)	0	2	M	2	4
Total telecommunications services offering score per country		15		9	24

Note: Area = area of availability, Price = price of telecommunications services

Area Price
+ = >80% L =the lowest 20%
o = 20%-80% M =the group in neither the highest nor lowest 20%
- = < 20% H =the highest 20%

munications services are added for each type of service depicted on one row and then the rows are added to get to the total telecommunications services offering score per country. We did not consider any weighting or further fine tuning of the starting values at this stage, although some international organizations might place different importance on several services. Weighting can be applied at a later stage after several cases have been done, so that a calibration can be made.

For each country, a matrix as in Table 4.3 can be developed. Multiple columns can be used to show several countries in one matrix.

Both the area of availability and price of telecommunications services and the telecommunications services offering score per country will be used in the "cost-effective management model" that will be described in Chapter 7. This model will be used to determine the relationship between the score and cost-effective management, using the input parameter values and output parameter values for each of the countries.

There are numerous sources of information to aid in filling in the telecommunications services offering matrix. As mentioned, in the case of a single telecommunications operator situation, that one operator is usually the only source. Any available publications are produced by the operator or by the government. When there are multiple telecommunications operators, often the regulatory body in the country can provide an overview. In most cases, however, we have to rely on other sources that provide the services of collecting the data, such as consulting organizations like Eurodata, Analysys and OECD [OECD 1997], as well as a combination of observations from the local offering brochures of telecommunications operators. Reports of consulting companies should be assembled quickly to obtain comparable results and time the measuring of the cost-effective management of a network with the measurement of the telecommunications services offering so the statistical analysis of the relationships will not be disturbed. The matrix can also be depicted in "condensed horizontal format" to describe multiple countries, as shown in Table 4.4.

4.5 SUMMARY

Most international organizations use various telecommunications services for their networks. A survey was done to determine telecommunications services network managers considered the most important, and their aspects. The set of available services in a country is called the telecommunications services offering.

Telecommunications services offerings may vary significantly between countries. Six services listed in the network managers' survey as most important were selected to become part of the telecommunications services offering matrix on one axis and the two aspects listed in the network managers survey as the most important (area of availability and price) were chosen to become part of the matrix on the other axis, so a comparison among countries can be made. The matrix will be used to compare the countries when the relationship between telecommunications services offering and cost-effective management is examined. The next step is to model the output parameter "cost-effective management" in a quantitative way, which we will do in Chapter 5.

TABLE 4.4
Telecommunications Services Offering Matrix in Condensed
Horizontal Format

Telecommunications Services Offering Matrix

	VPN				ISDN				PSTN			
	A	S	P	S	Total VPN	A	S	P	S	Total ISDN	A	S	P	S	Total PSTN
Country X	+	3	M	2	5	-	1	M	2	4	+	3	M	2	5
Country Y	0	2	M	2	4	+	3	H	1	4	+	3	H	1	4
Country Z	o	2	H	1	3	+	3	L	3	6	+	3	L	3	6

	Leased lines 64kbps					Leased lines 2 Mbps					PSDN			
	A	S	P	S	Total VPN	A	S	P	S	Total ISDN	A	S	P	S	Total PSTN
Country X	-	1	H	1	2	-	1	H	1	2	o	2	H	1	3
Country Y	+	3	M	2	5	+	3	M	2	5	+	3	M	2	5
Country Z	-	1	H	1	2	-	1	H	1	2	o	2	H	1	3

Totals

	Total Area	Total Price
Country X	11	9
Country Y	17	10
Country Z	12	10

Note: A = area of availability; S = Score; P = price of telecommunications services

5 Cost-Effective Management

In this chapter we will develop an approach to quantifying the output parameter "cost-effective management." We will address a methodology that can determine cost-effective management so that it can be quantified in any network. Section 5.1 gives an overview, and section 5.2 addresses cost relative to the size of the network. Section 5.3 addresses the performance of the network and how to measure that, and finally, section 5.4 shows the total calculation for cost-effective management.

5.1 OVERVIEW OF COST-EFFECTIVE MANAGEMENT

As explained in section 1.5, the research question asks for a qualitative relationship* between the input parameters and output parameter, but in order to establish a qualitative relationship, quantitative research is used, which will allow us to use statistical analysis to help assess qualitative relationships between the input parameters and output parameter.

Measuring the output parameter cost-effective management in a quantitative way is necessary for use in the statistical analysis. Cost-effective management can be calculated for the complete network or for a part of the network, e.g., the part in a particular country. As with the input parameters, the output parameter is used primarily on a per country basis in this research. The main problem with measuring cost-effective management is the lack of established methods to measure it. Most of the formulas in this chapter are based on taking several pieces of literature and combining them into one measurement methodology. Developing a methodology requires finding a balance between the accuracy of a measurement and the ease of application of the methodology.

To begin, a definition of cost-effective management should be established.

We define cost-effective management as *the degree to which the performance of the network fulfills the requirements of the users.*

---

* An example of a qualitative relationship: cost-effective management increases as costs of the telecommunications services used decrease.

Requirements of the users include performance indicators, size of the network, and costs of the network. Using the definition, cost-effective management can thus be split into the following principles, which we call cost-effectiveness principles (see Figure 5.1).

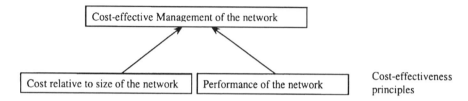

FIGURE 5.1 Cost-effectiveness principles as a basis for cost-effective management.

- Cost relative to size of the network
- Performance of the network (vs. required performance by the users in SLAs)

5.2 COST RELATIVE TO SIZE OF THE NETWORK

This cost-effectiveness principle, as the name suggests, is arrived at by comparing the *cost of the network* with the *size of the network*. We will analyze this, using, as much as possible, existing, universally applicable theories.

The cost of the network is calculated with the cost of ownership model of Treacy [Treacy 1989], described and refined in literature [Eekeren 1996, Kind 1996]. The size of the network is measured in terms of the number of nodes, links, or total capacity of the links. An explanation of the cost of ownership model is provided in section 5.2.1. Section 5.2.2 addresses the size of the network and section 5.2.3 relates cost to size.

5.2.1 THE COST OF OWNERSHIP MODEL

The cost of ownership model (COO model) was developed by Treacy at the Index Group [Treacy 1989, Looijen 1998] and has shown its value for doing calculations on costs of information systems and networks. Cost calculations can be useful in various situations, such as for requesting proposals from suppliers, comparing alternative deployment options, and budgeting network projects.

A few basic steps must be taken before the COO model can be applied. In particular, the demarcation of the network should be done. The demarcation can be a source of ambiguities if not properly defined before the cost of ownership model is applied. For instance, the nodes of the network contain equipment (such as computers, routers, and servers) that could either be considered part of the network or as peripheral equipment to the network. A personal computer (PC), for instance, contains a network adapter card, which could be considered part of the network, but the additional equipment in the PC, such as sound cards and hard disk, could be considered separate and not part of the network, depending on the chosen demarcation.

The cost of ownership model uses three types of costs:

1. Hardware and software
2. Personnel
3. External facilities

Hardware and software costs are lumped together into one type of costs, because in practice, there is a vanishing difference between hardware and software from a perspective of functionality. A particular functionality can often be implemented in either hardware or software. Hardware and software costs include cabling, network equipment, and servers. "Personnel costs" include administrative costs as well as salaries for personnel for network operations. "External facilities costs" include telecommunications services purchased from telecommunications operators, such as leased lines, as well as costs paid for other outsourced services. Hardware and software costs are split in the COO model into purchasing and maintenance costs, and purchasing costs are then spread over a certain period, which is usually 5 years in the form of depreciation.

When cost is calculated for individual countries, the cost of centrally located management that can be attributed to certain countries is allocated to them and added to the cost of the management in the country.

The cost of ownership model, shown in Table 5.1, is calculated using the cost data for hardware or software related to the network (includes depreciation and maintenance), personnel cost and actual cost of the external facilities, such as leased lines.

TABLE 5.1
Use of the Cost of Ownership Model

	Costs of Ownership (In thousands of US$ per year)			
	Hardware and software costs (1)	Personnel costs (2)	External facilities costs (3)	Total costs (1)+(2)+(3)
Country X	5	23	43	71
Country Y	4	32	23	59
Country Z	7	45	12	64

5.2.2 DETERMINING THE SIZE OF THE NETWORK

No standard method for determining the size of a network has been found in the literature. It is possible to use the number of nodes or the number of links or the total capacity of the links to determine the size of the network. In this research, we have chosen to determine the size of the network as the sum of total capacity of all links in the network or in the part of the network that is selected in a demarcation. The use of size in terms of capacity of the links, means that all services are included in that measurement. Some services use more capacity of a link than others, but the capacity used by a service is usually in relation to its functionality. For instance, a 2-Mbps leased line uses much more capacity than a 64-kbps leased line, but this is compensated by the fact that it also offers more functionality. For example, a 2-Mbps leased line can be used to carry more than 30 simultaneous phone conversations, whereas a 64-kbps leased line allows only one phone conversation at a time, although both numbers can be improved by using compression techniques..

5.2.3 DETERMINING THE COST RELATIVE TO SIZE OF THE NETWORK

The cost of ownership can be related to size by dividing the size of the network by the cost. In the example, this results in the overview of Table 5.2. It is also possible to relate the cost to size by using the reciprocal value, dividing cost by the size of the network. But in our total calculation for cost-effective management, the size-divided-by-cost representation of cost relative to size of the network is chosen because we prefer that a higher number indicates more cost-effective management and a lower number indicates less cost-effective management.

TABLE 5.2
Cost Relative to Size of the Network

Country	Size of the Network (1) (kbps)	Cost of Ownership (2) (US$/year)	Cost Relative to Size of the Network (1)/(2) (kbps/US$/year)
Country X	3456	71	48.7
Country Y	883	59	14.9
Country Z	2880	64	45

Tables 5.1 and 5.2 can also be combined into one table with multiple columns.

5.3   PERFORMANCE OF THE NETWORK

The second cost-effectiveness principle, performance of the network, is measured with performance indicators that are listed in the SLA or, if there is no SLA available, indicators that are measured and reported to the users [McGee 1996].

When an SLA is available, performance indicators are usually specified and tracked and the required values as specified by the users in the SLA as well.

Performance indicators that are associated with required values can be expressed in a percentage value that shows to what extent the performance indicator meets or exceeds the requirements in the SLA. If no required level is available, perhaps because there is no SLA, the performance indicator value itself can be used. It should be used so that a higher numeric value of the indicator means a higher level of performance. Therefore, if the performance indicator number is lower when a higher level of performance is measured, the reciprocal value of the performance indicator should be used instead. Such a performance indicator we will call an "inverse" performance indicator.

Research has been done on how to measure the performance of a PSTN [Cole 1991], usually by a regulatory body and often with help of a research institute, with the purpose of determining that the effects of competition in telecommunications stay within previously set boundaries. For example, the regulatory body monitors that calls to emergency services are not blocked more than an agreed percentage of the time.

Xavier [Xavier 1996] has shown several indicators that are being used in Australia. Examples are:

- Percentage of calls not completed on the PSTN
- Percentage of faults that are cleared within 24 hours
- Availability of itemized billing
- Quality of transmission of mobile telecommunications on the GSM network

User Surveys for Evaluation of Performance of the Network

One method of evaluating performance of the network is the user survey, which asks questions on performance of the network. Users can answer by showing their opinion on a scale, say, ranging from 1–5. An example of a survey in industry is Data Comm's annual user survey of international service providers [Heywood 1997]. This survey focuses on telecommunications operators and value-added service providers and performance indicators to meet their installation dates, minimize outages, and report and repair faults. The survey also asks users to rank the importance of the categories, so the weighted average result of the survey can actually give some idea of the level of satisfaction of the network service's users.

Indicators from Management Systems for Evaluation of Performance of the Network

Management systems are information systems that are used to support the activities of managing a network. Management systems can, for instance, keep track of the status of network elements and register when one becomes unavailable, a situation that is often indicated by the message "network element is down." As the system registers the time an element is down, it can also calculate the number of users affected and for what length of time. Registration of such information is a way to obtain performance indicators.

5.4 CALCULATING THE COST-EFFECTIVE
MANAGEMENT OF THE NETWORK

The approach described here for calculating the cost-effective management is shown in a chart depicted in Figure 5.2. Combining the numbers

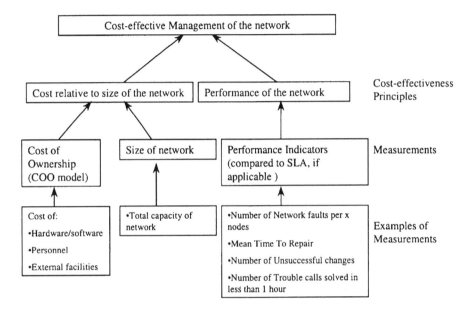

FIGURE 5.2 Measuring cost-effective management.

from the "cost relative to size" of the network calculation with the "performance of the network," we can assess cost-effective management of a
network. In the case studies of this research, cost-effective management
of a network is also measured on a per-country or a per-region basis.
Therefore, information on cost, size, and performance of the network
needs to be available on a per-country or per-region basis, depending on
how detailed the analysis of cost-effective management should be.

Since it is easiest to work with one number to express cost-effective
management in practice, a pro forma formula was invented:

$$\text{cost-effective management} \ = \ \frac{\text{performance of the network}}{\text{cost of ownership / size of network}} \qquad \text{(formula 5.1)}$$

However, care must be taken not to interpret the outcome of the formula
for cost-effective management as an absolute number that can be used to

compare networks, but rather as an indication of relative cost-effectiveness, given that all other circumstances are equal. With the use of a mathematical rule, formula 5.1 is equal to the formula for cost-effective management in simplified format (formula 5.2), which provides for easier calculation from the numbers in the earlier sections.

$$\text{cost-effective management} = \text{performance of the network} \qquad \text{(formula 5.2)}$$
$$\times \text{ size of network/cost of ownership}$$

When the SLA describes multiple performance indicators to measure performance of the network, multiple cost-effective management calculations will also be necessary. Due to the nature of performance indicators, they often cannot simply be added or combined into one single number. There may be different ways to combine different performance indicators in one number, but this is outside of the scope of this research, and if multiple performance indicators are used or requested by the users of the network, we will do separate calculations for each of the performance indicators.

5.5   SUMMARY

Although there is a lot of literature on cost-effective management and measuring performance of a network, there is little theory that provides actual measuring methods that are easy to apply. This chapter developed a formula for calculating an estimate of cost-effective management. Using an existing model for determining cost of a network, the COO model, cost of a network can be calculated and related to the size of the network. This is one of the two defined "cost-effectiveness principles" that together are defined to make up cost-effective management.

The other cost-effectiveness principle is "performance of the network," which is preferably expressed with the help of an SLA between the organization that manages the network and the users of the network. The SLA shows which performance indicator(s) should be used in the calculations for cost-effective management. The result is a series of numbers that indicate cost-effective management for the particular country or region that the numbers were based on.

In Chapter 6, step 6 of the research methodology will be covered. We will apply the input parameters "regulatory environment," and the "telecommunications services offering of the countries in which the network operates," and the output parameter "cost-effective management" in an explorative case.

6 Explorative Case

Step 6 of the research methodology is an "explorative case." After having modeled the input parameter "regulatory environment" in Chapter 3 and the "telecommunications services offering" in Chapter 4, and the output parameter "cost-effective management" in Chapter 5, an explorative case is carried out to determine how applying the models works in practice. The explorative case also provides a source of ideas for the cost-effective management model in step 7 of the research methodology and generates experiences in practice on what other international particularities managing an international network can involve. The chapter is organized as follows: Section 6.1 gives an overview of the explorative case for which the input parameters and output parameter are calculated and experiences with the calculation formulated. Section 6.2 shows the experiences in practice on international particularities that were encountered in the explorative case. Conclusions are drawn in section 6.3.

6.1 LUCENT TECHNOLOGIES

According to Yin, an explorative case study does not need to have a "proposition" but it does need a "purpose" and a "unit of analysis" [Yin 1984]. Here, the purpose is to discover what issues are important in practice, applying the the regulatory environment model and telecommunications services offering matrix of steps 3 and 4 as described in Chapters 3 and 4. The "unit of analysis" is a particular organization that was at the time of the explorative case study, a business unit of AT&T called AT&T Network systems. After 1996, the organization split from AT&T to become Lucent Technologies.

Lucent Technologies is an international telecom company with its headquarters in New Jersey. It employs about 120,000 people worldwide and has sales of US$30 billion annually.

The main products of AT&T are telecommunications services of all kinds. The main products of Lucent Technologies are telecommunications switching systems for public and private use, long-distance transmission systems, network management systems, wireless systems, microelectronic components, and consumer products such as telephones and answering machines. We will refer to the organization that we researched as Lucent Technologies.

The international network—a network spanning the world, with management represented in several countries—plays a role in processes for development, manufacturing, and sales of these products in the different countries in which the company operates.

Various services are being delivered over the international network, such as electronic mail, file transfer (including Internet and intranet traffic), and terminal session traffic.

The following subsections will give an overview of the network and an assessment of the entries in both models that we use as input for the cost-effective management model.

6.1.1 OVERVIEW OF THE NETWORK

The international network is operational in 42 countries. It covers the majority of the countries in which the company operates. The other countries may be connected by means of dial-up connections with modems over regular phone lines, or possibly not have data connection at all.

International Services Offering

The services provided by the Lucent Technologies international network are listed with the corresponding service value level in Table 6.1.

TABLE 6.1
International Services of the Lucent Technologies International Network

Service Value Level	Services	Locations
VALUE ADDED SERVICES	E-Mail	All locations
	Internet Access	All locations
BASIC DATA TRANSFER SERVICES	TCP/IP	All locations
	Asynchronous terminal session traffic	All locations
INFRASTRUCTURE SERVICES	Not available	

Layout of the Network

The network is built up in three layers that are called "backbone" (five nodes), "concentrator" (16 nodes) and "endpoints" (105 nodes). The nodes of the backbone are in Hong Kong; Allentown PA; Denver CO; Indian Hill IL; and

Hilversum (The Netherlands). The concentrator nodes are each connected to one of the backbone nodes as shown in Figure 6.1.

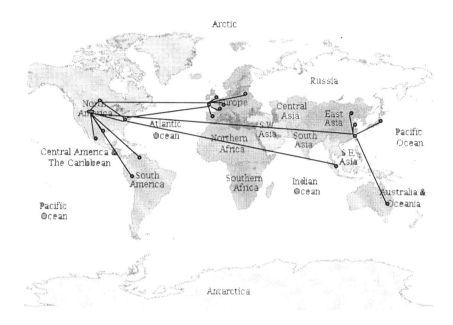

FIGURE 6.1 Backbone and concentrator nodes of the international network.

Table 6.2 gives a more detailed description of the backbone and concentrator nodes and the links between them.

Demarcation of Explorative Case Research

The backbone nodes handle the most traffic and have the most connected links in the network. As this is an explorative case, we do a first assessment using a limited number of countries. The three countries in which the backbone nodes lie are chosen, as in application of the regulatory environment model and telecommunications services offering matrix.

Evolution of the Network

Every description of a network is only momentary and is bound to change regularly [vanHemmen 1997]. We were struck by the fast evolution of the network in terms of capabilities, number of nodes, and type of equipment in the network. This resulted in the recommendation to assemble all information for the input parameters in a very short timeframe. As a reference

TABLE 6.2
Links Between the Concentrator and Backbone Nodes

Locations of the concentrator nodes	Connected with backbone node	Capacity (kbps)	Region
Sydney (Australia)	Hong Kong	128	Asia
Shanghai (China)	Hong Kong	64	Asia
Beijing (China)	Hong Kong	128	Asia
Singapore	Allentown	307	Asia
Tokyo (Japan)	Hong Kong	256	Asia
Munich (Germany)	Hilversum	64	Europe
Malmesbury (UK)	Hilversum	512	Europe
Hilversum (The Netherlands)	Allentown	1024	Europe
Nuremberg (Germany)	Hilversum	512	Europe
Plessis (France)	Hilversum	128	Europe
Bydgoszsz (Poland)	Hilversum	256	Europe
Madrid (Spain)	Hilversum	384	Europe
Reynosa (Mexico)	Denver	128	Americas
Lima (Peru)	Denver	64	Americas
Caracas (Venezuela)	Denver	64	Americas
Zafiro (Mexico)	Denver	128	Americas
Allentown-Indian Hill-Denver	3 ring	3072	Americas
TOTAL		8243	

Note: Americas = North, Central, and South America

for the impact of changes over a longer period, some important changes in the network from 1995–1997 are listed in Table 6.3.*

Management of the Network

Studying management of the network is a way to explore whether a cost-effective management indicator can be used. It can also be a way to learn some of the international particularities that are applicable to using an international network. Knowing international particularities can help in establishing checklists for use when an organization is expanding its network into a new country. Management of the international network is done from the network management centers, two of which have been

---

* Also, Lucent Technologies was spun off from AT&T in that period, but that event in itself did not influence the network being examined, because in the case of the year 1995, the network was separated into one business unit as well.

TABLE 6.3
Evolution of the Network between 1995 and 1996

Changes in the international network between June 1995 and January 1997

June 1995	January 1997
Not all international links encrypted	All international links encrypted
Backbone all leased lines	Backbone all Frame Relay connections
Voice and Data transported completely separately	Voice and Data integrated on the higher speed links. Lower speed links are still separate for voice and data services
Much asynchronous terminal traffic and little TCP/IP	Mostly TCP/IP services and little asynchronous traffic
Low volume data transfer over leased lines	Low volume data transfer over ISDN
Number of international links: 39	Number of international links: 67
Total number of links: 76	Total number of links: 114

established in Maitland FL and Hilversum (The Netherlands). A third center will be established in Singapore. The organization currently responsible for management of the network does not see it as a necessity to manage the network from the same location as where a backbone node is present. The choice for these locations was made on basis of availability of telecommunications services, availability of trained people, and locations in convenient time zones.

Three network management centres are used to adequately "follow the sun" [Ananthanpillai 1997], which means that the network most often has a group of telecommunications operators with a dayshift that can provide their full attention to problems. Apart from that, in order to maintain a high service level for the total network, right now the network management centers are still manned 24 hours a day with basic operations staff.

The follow-the-sun principle does not cover the world completely; Singapore is 5–6 hours ahead of continental Europe, depending on daylight savings time, and Maitland is 6 hours behind continental Europe, so there is a 2-hour overlap between business days. Some areas in the world, such as Hawaii, don't have the coverage of a fully staffed network management center during all of their business day, but have to operate 4 hours a day with a partly staffed network management center at their disposal. Follow-the-sun also requires that trouble tickets be handed off to the center located in the earlier time zone if the problems were not solved during the working day of the center that handled the trouble ticket.

6.1.2 ORGANIZATION OF NETWORK MANAGEMENT

The organization that manages the international network, the "global network services center (GNSC)," reports to the chief information officer (CIO) of the company and is structured as depicted in Figure 6.2.

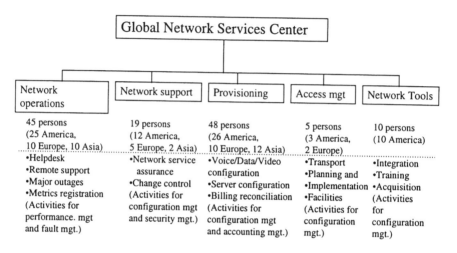

FIGURE 6.2 Organization of network management for the international network.

The functions of network management (configuration management, performance management, security management, accounting management and fault management) are all represented in the GNSC organization. Physically, the organization is spread around the world, exchanging information among the people in the organizations primarily by means of electronic mail.

6.1.3 REGULATORY ENVIRONMENT OF THE COUNTRIES

In this section, the model of the regulatory environment is applied to the situation of the explorative case of the international network.

The regulatory environment of the countries in the demarcation of the network is determined on the basis of publications, particularly in the journals on telecommunications policy* and the descriptions of the legal situations of countries in reports of consulting companies such as Analysys (http://www.analysys.co.uk) and Eurodata (www.eurodata.com). For our three countries, there is also a good source of information on the

---

* Journals such as *Telecommunications Policy* and *Global Telephony*.

Internet, which allows access to up-to-date information directly from consultants, the regulatory body or the ministry of communications. Table 6.4 gives an overview of the situation of the legal framework for competition in the selected centers.

TABLE 6.4
The Legal Framework for Competition Element
of the Regulatory Environment

	Legal Framework for competition						
	Value Added Services		Basic Data Transfer Services		Infrastructure Services		TOTAL Legal Framework for competition
	Code	Score	Code	Score	Code	Score	Total score
USA	C	3	C	3	C	3	9
Hong Kong	C	3	C	3	M	1	7
Netherlands	C	3	C	3	D	2	8

Table 6.5 gives an overview of the status of the regulations of the regulatory body of the selected countries. For the U.S., the sources for filling in the matrix were primarily the FCC website (http://www.fcc.gov) For Hong Kong, similar information was found on http://www.oftel.hk, and the information on The Netherlands came from http://www.opta.nl.

Table 6.6 gives an overview of the competition active element, with the appropriate scores. Counting the number of competitors is done with help of official publications of the regulatory body, or, if official publications are not available, using market research documents supplied by consultants.

All the input parameters in the models in this explorative case are current as of July 1996.

6.1.4 TELECOMMUNICATIONS SERVICES OFFERING
IN THE COUNTRIES

For assembly of the data, research by Eurodata and Analysys was used. The H, M, and L designations were chosen on the basis of comparison of a typical mix of usage and monthly subscription prices.

TABLE 6.5
The Regulatory Body Element of the Regulatory Environment

	Regulatory Body					
Categories (point scores)	USA		Hong Kong		The Netherlands	
1. *Independence / enforcement power*						
1a. Regulatory body independence	(2)		(2)		(1)	
1b. Does regulatory body have enforcement power?	Y (1)	3	N (0)	2	Y (1)	2
2. *Licensing process*						
Licensing process is:						
Short	N (0)		N (0)		Y (1)	
Transparent	Y (1)		Y (1)		Y (1)	
Nondiscriminatory	N (0)	1	N (0)	1	Y (1)	3
3. *Equal access to network infrastructure*						
A new telecommunications operator is allowed to:						
build network infrastructure	Y (1)		Y (1)		Y (1)	
have access to network of a dominant (bottleneck) telecommunications operator	Y (1)		Y (1)		Y (1)	
require mandatory co-location from a dominant telecommunications operator	Y (0)	2	Y (1)	3	N (0)	2
4. *Price of interconnection*						
Access charges are:						
cost-justified	Y (1)		Y (0)		Y (1)	
published	Y (1)		Y (1)		Y (1)	
non-discriminatory	Y (1)	3	Y (1)	2	Y (1)	3
5. *Fair competition*						
Price regulation	Price caps (1)		No (2)	Price caps (1)		
Universal Service						
	Y (1)	2	N (0)	2	Y (1)	2
6. *Number portability*						
Number portability mandatory	Y (3)	3	N (0)	0	N (0)	0
TOTAL Regulatory Body		14		10		12

TABLE 6.6
The Competition Active Element of the Regulatory Environment

| | Competition active | | | | | | |
| | Value Added Services | | Basic Data Transfer Services | | Infrastructure Services | | |
	Number	Score	Number	Score	Number	Score	TOTAL
USA	> 2	3	>2	3	> 2	3	9
Hong Kong	> 2	3	>2	3	0	0	6
Netherlands	> 2	3	0	0	0	0	3

Table 6.7, the telecommunications services offering matrix, shows the telecommunications services offering for the selected countries, spread over two tables.

Table 6.8 shows the totals of area and price of the telecommunications services offering matrix separately.

6.1.5 Total Scores for the Input Parameters

Now that all the input parameters for the chosen countries of this network have been filled out, the total scores of the countries can be calculated by adding all numbers per column and listing the totals on the bottom row. This row then shows the total score for each country for area of availability and price of telecommunications services, as well as the sum of the area of availability and price of telecommunications services. Table 6.9 shows a summary of all scores of the input parameters of the explorative case.

6.1.6 Cost-Effective Management of the Network

Knowing the number of people in each department and the activities of the department is essential as an input for the cost of ownership model to measure the output parameter cost-effective management.* Interviews with people in the various departments in the GNMC revealed details as to how many people were assigned to and sometimes physically working in a particular region. The cost of personnel was then calculated per region using US$60k as

---

* It turned out that activities of different people were not tracked. Detailed tracking of the activities of personnel can give a good idea of the cost of management activities.

TABLE 6.7
Telecommunications Services Offering Matrix

| | VPN | | | | ISDN | | | | PSTN | | | |
	A	S	P	S	Total VPN	A	S	P	S	Total ISDN	A	S	P	S	Total PSTN
USA	+	3	M	2	5	-	1	M	2	4	+	3	M	2	5
Hong Kong	0	2	M	2	4	+	3	H	1	4	+	3	H	1	4
The Netherlands	o	2	H	1	3	+	3	L	3	6	+	3	L	3	6

| | Leased Lines 64 Kbps | | | | | Leased Lines 2 Mbps | | | | | PSDN | | | | |
	A	S	P	S	Total	A	S	P	S	Total	A	S	P	S	Total PSDN
USA	+	3		3	6	+		L	3	6	+	3	L	3	6
Hong Kong	+	3	H	1	4	+	3	H	1	4	+	3	M	2	5
Netherlands	+	3	M	2	5	+	3	M	2	5	o	2	M	2	4

Notes: Area = Area of availability; Price = Price of telecommunications services; S = Score

Area	Price
+ = >80%	L = the lowest 20%
o = 20%-80%	M = the middle group
- = < 20%	H = the highest 20%

TABLE 6.8
Telecommunications Services
Offering Matrix Totals

	Total Area	Total Price
USA	16	13
Hong Kong	17	8
The Netherlands	16	13

average cost per person per year. All information is current as of the date of the case study (July 1996).

The total cost of ownership, shown in Table 6.10, was calculated using actual cost data for hardware/software related to the network (includes

TABLE 6.9
Summary of the Scores of the Input Parameters

	Input Parameters						
	Legal framework for competition (1)	Regulatory body (2)	Competition active (3)	Total regulatory environment (1) + (2) + (3)	Area of availability (4)	Price (5)	Total Telecommunications Services Offering (4) + (5)
USA	9	16	9	34	16	15	31
Hong Kong	7	9	7	23	17	8	25
Netherlands	8	12	5	25	16	13	29

TABLE 6.10
Cost of Ownership

	Costs of Ownership (In thousands of US$ per year)			
	Hardware-software costs (1)	Personnel costs (2)	External facilities costs (3)	Total costs (1)+(2)+(3)
USA	3200	4560	192	7952
Hong Kong	1500	1440	1324	4264
The Netherlands	2300	1620	3168	7088

depreciation and maintenance) and actual cost of the external facilities (leased lines).

The size of the network was determined, and cost relative to size of the network was calculated by dividing the size by the cost as explained in section 5.2. This results in the overview of Table 6.11.

After calculating cost relative to size of the network, the formula for cost-effective management describes a calculation with performance indicators, which were difficult to choose. In total, five performance indicators were tracked by the network management organization. The indicators were:

TABLE 6.11
Cost Relative to Size of the Network

	Size of the network (1) (kbps)	Cost of Ownership (2) (kUS$/year)	Cost relative to size of the network (1)/(2) (kbps/kUS$/year)
USA	3456	7952	0.436
Hong Kong	883	4264	0.207
The Netherlands	2880	7088	0.406

- Network faults per 1000 nodes
- MTTR, rate of unsuccessful changes in the network relative to the total number of changes
- On-time completion of provisioning
- Utilization of the network

A problem was that these performance indicators were not available for the regions or countries individually, but were only tracked for the complete network. Also, there was no SLA between the network management organization and the users of the network. Due to the lack of an SLA, it was not certain that the five performance indicators that were tracked were actually the indicators that are most important for the users of the network. This leads to the conclusion that it is necessary to track the indicators regularly on a per-country or per-region basis. If there is no SLA, at least a confirmation should be obtained that the tracked indicators are relevant for the users. This can be done by means of a survey.

6.2   INTERNATIONAL PARTICULARITIES IN PRACTICE

As addressed in section 1.4, international particularities found in the literature [Liebmann 1995] were of various natures and many of them were not exclusively tied to a country. They did, however, provoke interest in other particularities. Furthermore, knowledge on international particularities is essential for a network manager that has the task of expanding a network internationally. They could, for example, be included in a checklist that an international organization might use for the expansion of a network into a new country. The explorative case gives a practical situation in which international particularities are found. In this section, interviews with the people who manage the network were used to generate the list of international particularities, which are organized using the functional areas of network management.

International Particularities and Functional Areas
of Network Management

This section uses the activities of the "system management functional areas," a part of the OSI management framework that is explained in section 1.7.3.

There are several different interpretations of what activities are part of the functional areas. We use a description by Terplan [1992] to clarify the activities of the functional areas. The particularities differ also in their reason for inclusion. Four categories are:

1. Country
2. Language
3. Telecommunications operator
4. Other reason

In brackets, the category of particularity is denoted by a "C" (reason for particularity is the different country) if it is actually dependent on the country. Two other categories depicted here are "L" (reason for particularity is the different language), "T" (reason for particularity is the different Telecommunications operator) or "O" (reason for particularity is caused by other differences).

Configuration management—carries out middle- and long-range activities for controlling physical, electrical, and logical inventories, maintaining vendor files and trouble tickets, supporting provisioning and order processing, defining and supervising SLAs, monitoring ongoing changes and distributing software.

International particularities for configuration management

- SLAs are governed by different legal regulations in the countries (C). This is country dependent, because legal regulations are usually applicable for one particular country.
- Some services to be delivered to other parties or to the real system (arrows B and C in the network management paradigm; see section 1.7) cannot be of adequate quality and should therefore not be performed. Example: They provide only store-and-forward services in countries with lower quality basic data transfer services (T). This is dependent on the operator in the country that offers the services.
- Internet domain name (for electronic mail) addresses should be standardized (C); every country has a standard for Internet domain

name addresses, usually supervised by an Internet domain name supervising body.

Fault management—a collection of tasks required to dynamically maintain the network service levels. These tasks ensure high availability of computing resources by pinpointing problems and performance degradation quickly and, when necessary, by initiating controlling functions that could include diagnosis, repair, testing, recovery, and backup.

International particularities for fault management

- Information for surveillance of the status of a link should be carried on a link separate from the one that transports the information (T); this is dependent on the operator, as different operators will have different policies on transport of information for surveillance of status.
- Information regarding the status of the network, including "trouble tickets," should be multilingual in some countries (L), depending on what languages are commonly used in the country, or what language(s) the international organization has chosen to use.
- Help desks should be available during the user's working hours [Bishop 1996], because of time zone differences (O) and the celebration of different holidays in different countries (C); this depends on the country in case of holidays and mostly on the country for time zones, however, in some countries, multiple time zones exist. Examples of countries with multiple time zones are: Russia with 11 time zones and the United States with four.
- Help desk employees should have the language skills to support everybody in the real system (L); this depends on the languages that are used in the real system.
- Although standard equipment might be used throughout the network, suppliers don't always support a particular type of equipment in every country (C); equipment suppliers usually choose a country as an area for marketing, aiming their support organizations and advertising for use throughout a country, and usually not across international borders.

Performance management—a set of activities required to continuously evaluate the principal performance indicators of network operation, to verify how service levels are maintained, to identify actual and potential bottle-necks, and to establish and report on trends for management decision making and planning.

International particularities for performance management
Use of performance indicators should be defined so that performance can be measured across borders (T); performance indicators are related to the operator(s) providing the service. Performance indicators might be:

- Availability of agreed bandwidth
- Quality of the connection (also called "line transmission quality")
- Average outage time (also called MTTR)

Compression effects come into play as international links are compressed because of high costs. This introduces extra *propagation delay* in the transport of information that can have a negative effect on certain applications. For example, propagation delay for applications such as terminal sessions and voice transmission should be below 150 milliseconds, so compression possibilities are limited for those applications (T); it is common to compress information on international links and not to do so on domestic links, but it is entirely a decision of the operator or group of operators that are providing the service.

Security management—the ongoing protection of network and its components—such as the protection against entering the network, accessing an application, and transferring information in a network. It includes:

- protecting network management tools by analyzing risks and minimizing risks
- developing and implementing a network security plan
- monitoring results of the network security plan

International particularities for security management

- When multiple telecommunications operators are suppliers of services, extra risks of fraudulent entry are introduced into the international network (T); this depends on which telecommunications operator delivers the services and whether it uses multiple operators at a lower level in the service value model.
- Different legal systems in countries have different possibilities for prosecuting (and thus deterring) criminal activities on an interna-

tional network (C); this depends on the country, as laws are usually applicable in one country only. In some countries, slight differences in laws exist between states or provinces, but if there are laws governing the security of an international network, most countries' laws are applicable across the whole country.

- Determination in which country the crime has been committed is important for criminal law, but may be difficult due to the international nature of the network. In most countries, criminal law states that, in principle, the law is applicable only to the country in which the criminal activity has been conducted. If the physical location of the activity is not clear, this introduces extra difficulties in pursuing legal action. It may even be possible that the activity was conducted from a country where the activity is not considered illegal (C); this depends on the country in question, as it is tied to the laws of a country.

- Encryption of information is illegal in certain countries. For reasons of control, some countries prohibit the sending of encrypted information across networks (C); this depends on the country in question, as this is tied to the laws of that country.

- Encryption algorithms cannot be exported to certain countries. In particular, the U.S. Department of Defense has mandated that only certain types of encryption algorithms may be exported outside of the United States. Depending on the country to which the encryption algorithm is, there may be have a certain level of security against cracking (C); this depends on the country to which the encryption algorithms are exported.

Accounting management—the process of collecting, interpreting, and reporting costing and charging-oriented information on resource usage. In particular, processing of accounting records, bill verification, and charge-back procedures

International particularities for accounting management

- Identification of cost components is necessary per country for fiscal authorities of the country. Local assets that are part of the international network should be booked on local balance sheets and on local income statements (C); this depends on the accounting principles that are applicable in the country and on the tax system of that country, which is usually governed by laws and therefore specific for the country.

- Splitting of bills must be done with the correct local currency for each of the countries. This is also for the identification of the cost components as mentioned in the previous point (C); this depends on the country, as countries usually use their own unique currency, although some do use a common currency.

Conclusions on International Particularities

The explorative case shows that international particularities do exist, although some are not just caused by a link crossing an international border, but might also exist regardless of the existence of a country border. Many of the international particularities have only been discovered recently, as networks have started to become more international, as explained in the trends in section 1.2. As any of these (and more) might be encountered when expanding a network internationally, attention should be given to these international particularities, and procedures should be developed to handle them.

A few observations from the study of international particularities in this explorative case are:

- International particularities that are dependent on language come into play in particular in fault management activities.
- Performance management particularities are mostly caused by the differences in telecommunications operators.

Since this is data based on an explorative case and combined with literature research, conclusions might not be universally applicable, but might, for instance, be used as a basis for development of a checklist that could be important to review and that could support the expansion of a network into a new country.

6.3 SUMMARY AND CONCLUSIONS FROM THE EXPLORATIVE CASE

The regulatory environment, telecommunications services offering, and the formula for cost-effective management, described in Chapters 3, 4, and 5, were applied in the international network of AT&T Network systems/Lucent Technologies. The explorative case shows that applying and entering numbers in the regulatory environment model and the telecommunications services offering matrix is generally feasible, but in specific situations, the data might not be available.

Because of the rapid evolution of the network, the recommendation was made to gather information for filling out the regulatory environment model and the telecommunications services offering matrix in a very short time window. The cost-effective management (output parameter) was only in part calculated. Although capacity and cost information were available on a per-country or per-region basis, the performance indicators were not tracked on a country- or region-specific basis and the cost-effective management calculation could therefore not be fully completed. This led to recommendations to improve the measurement of performance indicators. Another conclusion is that the use of time by employees of the organization in question be tracked per activity of support people in the network management department and that the tracking of indicators be studied to ascertain tracking of country-specific performance information. The next step in the research methodology, step 7, which concerns the development of a cost-effective management model, is described in Chapter 7.

7 The Cost-Effective Management Model

Step 7 of the research methodology is the development of a cost-effective management model, which is addressed in this chapter. First, however, the position of a cost-effective management model should be clear in terms of the process theory. Having determined input and output in steps 3 through 5 of the research methodology, we consider the process, which takes place between input and output, as depicted in Figure 7.1.

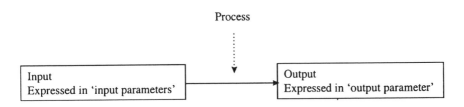

FIGURE 7.1 Cost-effective management model: determining the process between input and output

The process between input and output, expressed respectively with input parameters and an output parameter is considered in terms of relationships, as mentioned in the research question. Those relationships and their expression in a cost-effective management model are the main subject of this chapter

Section 7.1 addresses the requirements for a cost-effective management model, following which we will show a proposal for a cost-effective management model and discuss the various ways to test it, based on propositions in section 7.2. Section 7.3 explains propositions with help of heuristics, and section 7.4 addresses a possible testing method through statistical analysis.

7.1 REQUIREMENTS FOR A COST-EFFECTIVE MANAGEMENT MODEL

There are various requirements for a cost-effective management model. Apart from the fact that it should help answer the research question and thus help to show the relationships between input parameters and output parameter, the model should also be easy to apply. Applying it should require only a reasonable amount of time, so that it can adequately help support the answering of the strategic questions mentioned in section 1.3 when applied on a daily basis in practical situations.

The regulatory environment model, combined with the telecommunications services offering matrix form the input parameters of the cost-effective management model. Its output parameter is cost-effective management. Both the input parameters and the output parameter have been modeled and described. Figure 7.2 depicts the input parameters, process, and output parameter in a conceptual view, which, with relationships added that detail the process, would form the cost-effective management model.

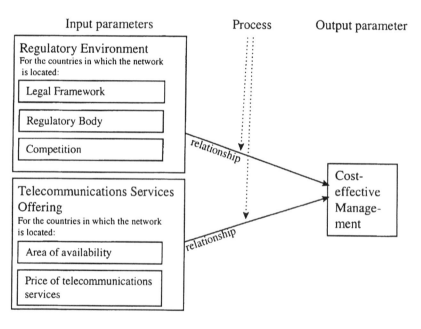

FIGURE 7.2 Conceptual view of the input parameters, process, and output parameter.

The arrows marked "relationship" show the relationships we are seeking to describe in order to answer the research question. However, to learn more detail of the relationships, we must consider the use of propositions. A proposition is described in the literature as *a statement about the causal connections between abstract concepts* [Knoke 1991]. The abstract concepts translate into the input parameters and the output parameter. The causal connections are the relationships.

7.2 PROPOSITIONS FOR THE COST-EFFECTIVE MANAGEMENT MODEL

As explained, propositions are statements about the relationships between input parameters and the output parameter. To develop propositions, we analyze the input parameters. Models for the input parameters regulatory environment and telecommunications services offering have been developed in steps 3, 4, and 5 of the research methodology so that they represent higher scores when one of the following situations is applicable in a country, as supported by the description of the regulatory environment in section 3.4.

- Legal framework for competition shows a relatively high number of service value levels in which competition is allowed.
- Regulatory body shows a relatively high number of categories with high scores, which means that relatively many measures to promote good competition have been implemented.
- Competition active shows a relatively high number of competitors actually competing with services in various service value levels.

The literature [Noam 1997, Frieden 1995] suggests that when the situations mentioned above are applicable in a country and thus the score for the regulatory environment is high, good conditions have been created for organizations to manage their networks in that country. Cost-effective management has been defined so that a relatively high score for it would represent one or more of the following situations.

- The cost of ownership of the network is relatively low.
- The size of the network is relatively large.
- The performance of the network is relatively high.

As a result, we can formulate a proposition that ties a high score for the regulatory environment to a high score for cost-effective management.

Proposition 1: An increase in the score of the Regulatory Environment in a country results in an increase in cost-effective management of the network in that country.

A similar approach can be taken for the input parameter telecommunications services offering, as described in section 4.4. A high score in the telecommunications services offering matrix indicates one of the following situations:

- *Area of availability* shows a relatively large area of the country where many telecommunications services are available.
- *Price of telecommunications services* shows a relatively low price for telecommunications services in the country.

No literature has been found that suggests the influence of the telecommunications services offering on cost-effective management, but the presence of "cost of external facilities" in the cost of ownership model generates a suggestion. The explorative case suggests that a considerable part of the cost of external facilities is made up of telecommunications services purchased from telecommunications operators. When "price of telecommunications services" increases, which is shown by a lower score in the telecommunications services offering matrix, the "cost of external facilities" increases. This results in a higher "cost of ownership," which in turn means a lower score for cost-effective management, as supported in section 5.4. Likewise, an increasing score for price of telecommunications services could result in an increasing score for cost-effective management.

Following is a proposition that ties an increase in the score for the telecommunications services offering to an increase in score for cost-effective management.

Proposition 2: An increase in the score of the telecommunications services offering in a country results in an increase in cost-effective management of the network in that country.

Using the Propositions for Further Research

We will explore two methods to assess the propositions. The first is by using heuristics, here called "heuristic intermediate factors" to help explain the relationship. Heuristic intermediate factors (HIFs) do not use calculations or process quantified information, but rather use reasoning and relations obtained from literature. This method is addressed in section 7.3.

The second method to learn about the relationships is to do a statistical analysis as shown in section 7.4. A statistical analysis uses quantified information from the input parameters as calculated in the regulatory environment model and the telecommunications services offering matrix of case studies.

7.3 PROPOSITIONS EXPLAINED WITH HEURISTICS

Each of the propositions shown in section 7.2 is examined and possibly explained with heuristics. This means that heuristics are used to find a reasoning that confirms the relationship. For Proposition 1, HIFs are used as aids. These are ideas that help bridge the mental gap between input parameters and the output parameter when thinking about the relationships.

7.3.1 EXPLANATION OF PROPOSITION 1

Proposition 1 relates the regulatory environment in a country to cost-effective management. This relationship can be explained by introducing two HIFs. The first shows a relationship between the regulatory environment and innovation, as well as between innovation and cost-effective management. HIF1 is expressed as follows.

Regulatory Environment *relates to* innovation and innovation *relates to* cost-effective management.

HIF1 assumes that regulatory environment relates to innovation, which relates to cost-effective management. Both regulatory environment and cost-effective management are terms that are defined, but innovation should still be defined. HIF 1 is formulated from two statements, one concerning the relationship between regulatory environment and innovation and one concerning the relationship between innovation and cost-effective management.

As stated in Chapter 3, the score of the regulatory environment is higher if there is more competition allowed. The relationship has been shown by van Cuilenburg and Slaa [1995] to be: *the more competition allowed, the greater the innovation.* The literature makes a distinction between various kinds of innovation, and mentions in particular "product innovation," which means a greater choice of services.* This shows the likelihood of the first half of HIF1.

As more innovation means more choice of services, a lower cost of managing the network will follow as more services become available to choose from. The cost of services purchased for the network can be found in the COO model under "external facilities" and are shown in the explorative case to be a considerable part of the cost of ownership of an international network. Lower cost in the COO model result in higher scores for cost-effective management, according to the formula for cost-effective management in section 5.4.

---

* Van Cuilenburg and Slaa examine two different kinds of innovation: Process innovation is "more of the same" (just a lower price); product innovation means the creation of new products that did not exist before.

A second reason to explain HIF is a researched relationship between the regulatory environment and a performance indicator, in literature often named "quality of service." To explain this reason, we need a measurement of quality of service [Cole 1991, Frieden 1996]. For example: a few common performance indicators for PSTN services are system response time or speed of dial tone, the number of trouble reports compared with the number of users, and call completion. The indicators differ according to the type of service. The overviews of Frieden and of Cole show that three to four years after a change in the regulatory environment that resulted in more competitors for each of the telecommunications services, there was a measurable improvement in quality of service.* These indicators are used for investigation of a relationship between the regulatory environment and the performance of the network. A higher performance of the network filled in using the formula of cost-effective management of section 5.4 means that cost-effective management of the network increases.

To create an explanation that is as thorough as possible, a second HIF was generated using price of telecommunications services. HIF2 can be stated as follows.

Regulatory environment *relates to* price of telecommunications services and price of telecommunications services *relates to* cost-effective management.

HIF 2 concerns the relationship between the regulatory environment and price of services as well as the relationship between price of telecommunications services and cost-effective management. Research by the OECD [1995] compared several countries' telecommunications services prices and showed that more competition (which is a higher score in the regulatory environment model) means lower prices for telecommunications services. Figure 7.3 shows an overview of the changes in prices for telecommunications services in 15 OECD countries between 1990 and 1997. Prices in competitive environments turn out to be lower than in noncompetitive environments.

Lower prices of telecommunications services can result in a higher score of cost-effective management, because prices of telecommunications services are represented in the cost of ownership model and therefore in the calculation of cost-effective management.

---

* The performance indicator "quality of service" in Cole's study is expressed in transmission quality, average PSTN dial tone delay, percentage of service orders completed on time, percentage of calls completed, and customer perception surveys [Cole 1991 p. 261].

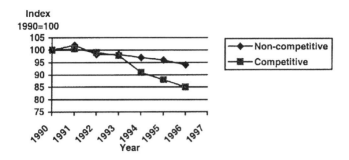

FIGURE 7.3 Comparison of telecommunications prices in competitive and noncompetitive environments [OECD 1997].

7.3.2 Explanation of Proposition 2

Proposition 2 relates the telecommunications services offering in a country to cost-effective management. Proposition 2 can be split into two statements, each forming one half of the proposition. One relates to area of availability and one to price of telecommunications services. The relationships of these two statements are explained in this section. The explanation can be made without using HIFs.

1. *Area of availability relates to cost-effective management.* The area of availability of services is expected to correlate with the cost-effective management of the international network. The type of influence is dependent on what type of service the network should perform for its users. To explain the expected influence, the network is viewed with the services it performs and the services it purchases in the value chain [Porter 1998], as explained in section 1.7.2 and again in Figure 7.4.

FIGURE 7.4 Telecommunications services in the international network value chain.

Table 7.1 shows an example of how availability of services to be purchased might influence cost-effective management. The middle and left columns show services that can be purchased to perform services in the right column. The basic data transfer service shown in the left column is the least

TABLE 7.1
Example of Influence of Area of Availability on Cost-Effective
Management

Services purchased		Services performed
More cost effective when these basic data transfer services are available for purchase. (less-expensive services)	Less cost effective when these basic data transfer services are available for purchase (more-expensive services)	
• Leased lines (if volume of voice services is high)	• Public Switched Telecom Network (PSTN)	• Voice services between nodes in a network
• Packet Switched Data Network (PSDN)	• Leased lines	• Terminal connections
• ISDN or Frame Relay	• Leased lines	• File Transfer Services
• Frame Relay or ATM	• Leased lines	• High volume file transfer

expensive service available to perform the value added service in the right column. If that basic data transfer service is not available in a part of the country, a more expensive service, such as shown in the middle column, must be purchased.

With the help of overviews such as Table 7.1, it is possible to improve cost-effective management of the network by selecting the service to be purchased for the delivery of a service by the international organization.

2. *Price of telecommunications services relates to cost-effective management.* The second half of Proposition 2 concerns the relationship between price of telecommunications services and cost-effective management. The assumption that price of telecommunications services has influence on cost-effective management of the network is shown in the definition of cost-effective management and the cost of ownership model in section 5.4.

7.3.3 TOTAL EXPLANATION WITH HEURISTICS

The cost-effective management model with HIFs is shown in Figure 7.5. A dotted arrow shows the expected correlation. A complete relationship between input parameter, HIF, and output parameter is identified by HIF 1 and HIF2.

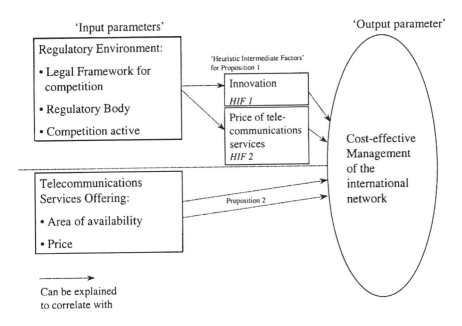

FIGURE 7.5 The cost-effective management model.

Correlation can be a positive relationship or a inverse relationship. In other words, when, for example, innovation (HIF 1) increases, cost-effective management increases, denoting a *positive* relationship.

It is possible to explain the propositions with heuristics, which, however, can by definition not accept a proposition as true; there may be other situations that yield results that could contradict the assumed propositions.

Heuristics, however, may lead us to a better understanding of what can influence cost-effective management, and therefore can show relationships. The statistical research makes it possible to analyze the individual elements of the input parameter and help understand possible relationships between input parameters and the output parameter, however, we should realize that the values of the input parameter are nominal values.

7.4 PROPOSITIONS TESTED WITH STATISTICS IN THE CASES

The second method to test the propositions is a statistical analysis of the relationships between the regulatory environment model and the telecommunications services offering matrix on one hand, and the cost-effective management of the network on the other. The method using statistics also goes a step further than the method with heuristics, since it looks at the individual elements of the regulatory framework and telecommunications services offering.

A statistical method that assesses the relationship between two series of data points is called "regression analysis." Regression analysis shows the likelihood of a relationship between input variable(s) and an output variable. Regression analysis can be applied with single input variables or with multiple input variables, but the number of calculations increases rapidly when many input variables are used. The result of regression analysis is a series of statistical indicators that give an indication of how the input parameters and output parameter could be related. Statistical analysis does not result in statements with 100% certainty about the relations and, as with heuristics, cannot be used to accept the "propositions" of section 7.2. Regression analysis does, however, give a quantified assessment of the relationship. Chapter 8 will show a detailed overview of the approach of the cases.

7.5 SUMMARY

The cost-effective management model as developed in this chapter intends to describe the process that takes place between input and output, exemplified by the input parameters and output parameter respectively. The process is expressed in relationships between the input parameters and output parameter.

Two propositions were developed that show assumptions about the relationships. Two methods are proposed to assess the validity of the propositions. One used heuristics, which gave a high-level explanation of why the propositions might be applicable. For the high-level explanation, "heuristic intermediate factors" were used. The first method strengthened the confidence in the propositions. This led to the decision to analyze the models in further detail and do case studies in the second method.

8 Description of the Test Cases

This chapter, step 8 of the research methodology consists of carrying out two test cases. The purpose of the test cases, hereafter called "the cases" is to test the heuristically found qualitative relationships between the input parameter "regulatory environment" in the country and the country's telecommunications services offering, and the output parameter cost-effective management of the network. This chapter will first address the requirements for the cases in section 8.1, after which the two test cases follow in sections 8.2 and 8.3.

As the propositions call for an examination of the scores on a per-country basis, one of the requirements for the cases will be that the input parameters and output parameter are discernable on this basis. Further requirements are developed in section 8.1.

8.1 REQUIREMENTS, ACTIVITIES, AND DESCRIPTION OF THE CASES

The following four requirements for cases were decided on before the cases were chosen.

1. Cases should concern an organization that operates in a different industry.
2. Cases should involve a several countries, so that the statistical analysis can be done with confidence. A minimum number of five countries is assumed for each of our cases.
3. Cases should have a stable network environment that does not change drastically within the time frame of observation. As many measurements are taken over a period of time, rather than as a snapshot, it is necessary that there be no sweeping changes in network or the managing organization in the observation period.
4. Cases should operate in countries where data on the input parameters *regulatory environment* and *telecommunications services offering* are available.

Two organizations were chosen that fulfill the requirements and, in addition, were willing to provide the detailed information needed for a case study.

1. The Calvin Klein Cosmetics Company (CKCC) is a commercial organization in the cosmetics industry that manufactures, markets, and sells cosmetics. CKCC maintains branches in seven countries.
2. The United Nations Development Programme is an international governmental organization maintains bases in 143 countries.

See Table 8.1 for an overview of how both case organizations fulfill the requirements.

TABLE 8.1
Case Organizations and Fulfillment of Requirements

Requirement	CKCC	UNDP
Industry	Commercial organization in cosmetics industry	International governmental organization
Number of countries of organization	7	143
Stable network environment	Yes	Yes
Operates in countries where data is available for input parameters	Yes	Yes

With the case study organizations selected, we start the process of carrying out the case study and then describing it. To this end, a series of activities that support a systematic approach was established (see Figure 8.1).

The people interviewed were network managers and managers of the information technology departments. Experience has shown that several interviews are necessary to collect adequate information, and often the information about performance of the network must be calculated from system reports.

The information from both a questionnaire and the interviews is then combined with separate research on country data and entered in the regulatory environment model, the telecommunications services offering matrix, and the cost of ownership model and the scores were analyzed with statistics.

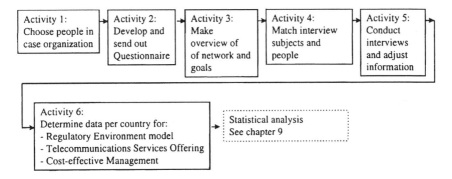

FIGURE 8.1 Flow of the activities for the cases.

Activity 1

Choose interviewees in the case organization to provide us with information. A start can be to contact the chief information officer (CIO) of the organization or the person responsible for management of the network. That person can then identify other people to provide information.

Activity 2

Develop and send out a questionnaire (see Table 8.2) to precede the interviews with the interviewees to help lay the groundwork for the interview process. The questionnaire not only aids the interviewer, but will help make the selected people in the organization aware of what questions will be asked of them and should prompt them to assemble the right information, reports, or colleagues to be prepared to furnish in an interview any pertinent information that the questionnaire might have missed.

Activity 3

Create an overview of the network and goals of the organization. This is done to understand the organization better, so the importance of the network to attaining the organization's goals can be judged, and to assemble information about the network that will be needed in the calculation of cost-effective management. This activity also should include a demarcation of the network, in case it cannot be studied in its entirety.

TABLE 8.2
Sample Questionnaire and Purpose of its Questions

Question	Purpose
What is the primary process of the international organization?	Used to determine if performance indicators used in the network are relevant.
Which services does the network deliver/what applications does it support?	Used to check if cases are comparable.
In which countries is the international network located?	Used as basis for the scoring of the regulatory environment model and telecommunications services offering matrix.
What is the topology (links, nodes, capacities)?	Used to determine the size of the network; essential for cost-effective management.
What is the content of the SLA or any other established objectives? (indicators and benchmark values) if applicable	Used to determine the reference levels of the indicators.
To what extent are the performance indicators in the SLA being attained? (What are values of the performance indicators?)	Used to calculate cost-effective management.

Activity 4

Match interview subject and people chosen to be interviewed. The preparation helps to ascertain that the right people will be interviewed and that each one is asked the questions that pertain to their position in the organization.

Activity 5

Conduct interviews and review information gathered. The interviews should elicit all the required information on the network and management. Using the interviews, go back to the questionnaires and fill out any incomplete questions. Assemble quantitative information of the case study for the calculations in Activity 6.

Activity 6

Determine the data per country for required to fill in the regulatory environment model and the telecommunications services offering matrix, and for the calculation of cost-effective management.

The regulatory environment model and telecommunications services offering matrix are filled in with data from consultant reports, or, if not available, by contacting a regulatory body or government organization in the countries. The calculations are then made as described in Chapters 3 and 4.

To calculate cost-effective management, the model described in section 5.4 will be applied in the case study, using Formula 5.2.

Cost of ownership, size of the network, and performance of the network are the measurements that must be assembled for the cases. For each of the countries, cost of ownership is calculated using cost information obtained in the questionnaire and through interviews. The size of the network is measured by taking the totals of the capacity of the links for each of the countries in the case. As described in section 5.3, performance of the network consists of performance indicators that are preferably based on the content of an SLA, or, if this is not present, requirements expressed by users in the interviews can also be used.

After the six activities are carried out, the case is described and the data can be analyzed statistically, which will be done in Chapter 9.

In the following sections, the cases will be described, using the following structure.

1. Overview of the network.
   ♦ International services offering, including services offered to the users of the network and the locations
   ♦ Layout of the network shown in a graphical format with the capacities of the links
   ♦ Demarcation of the case research, including a list of countries chosen for analysis
2. Organization of network management showing primarily the size and location of the organization responsible for managing the network.
3. Regulatory environment of the countries showing the score in the regulatory environment model.
4. Telecommunications services offering of the countries showing the offering in the matrix.
5. Total scores for the input parameters, showing a summary of the scores for the regulatory environment and the telecommunications services offering.

6. Cost-effective management of the network, including
   ◆ Cost of ownership
   ◆ Cost relative to size of the network
   ◆ Performance of the network
   ◆ Cost-effective management calculations

8.2   CASE 1: CALVIN KLEIN COSMETICS COMPANY (CKCC)

Calvin Klein Cosmetics Company is a 100% daughter company of Unilever North America and manufactures, distributes, and sells cosmetics, fragrances, and personal-care products. It is co-marketed with the clothing brand of Calvin Klein all over the world. The organization for manufacturing and distribution is located in North America and Europe, and maintains an international network with about 1000 users.

8.2.1 OVERVIEW OF THE NETWORK

International Services Offering

The CKCC international network primarily performs data services. In particular, the connections between the U.S. headquarters and the affiliate offices in Europe carry data exclusively. A number of value added services, such as electronic mail applications (cc Mail), groupware applications (Lotus Notes) and both Internet and intranet access through the backbone network of the parent company Unilever, are available on the network. Typically, the network is also used for exchanging transactional information in electronic data interchange (EDI). At the basic data transfer services value level, the services performed for the users are TCP/IP* file transfer and terminal session connections over leased lines.

The network consists exclusively of leased lines and ISDN connections used as a backup to the leased lines. The leased line bandwidth is generally 64 Kbps, with the exception of the connections to and from the U.S. to Germany and U.S. to Canada which have a bandwidth of 56 Kbps.

The only higher capacity link in the network is the T-1 (1.544 Mbps) connections between Mount Olive and New York. Table 8.3 presents the services used on CKCC's international network in the service value model

---

* TCP/IP (Transaction Control Protocol/Internet Protocol) is a packet switched data protocol, commonly used in the Internet and intranets.

TABLE 8.3
Services on the International Network of the Calvin Klein Cosmetics Company

Service Value Level	Services	Locations
VALUE ADDED SERVICES	Voice Public Switched Telecom Network (PSTN)	All locations
	e-Mail	All locations
	Internet Access	All locations
	International Integrated Services Digital Network (ISDN)	All locations
BASIC DATA TRANSFER SERVICES	TCP/IP	All locations
	International leased lines:	All locations
INFRASTRUCTURE SERVICES		Not available

(services in the middle column and the locations where they are available in the right column).

Layout of the CKCC International Network

The company's international network covers North America and Europe. The main hub is located in Paris, where Calvin Klein facilities are located on the premises of France Télécom. The network backbone consists of two 64 Kbps leased lines between the main hub in France and a smaller hub in Mount Olive, NJ.

The Paris hub (see Figure 8.2) establishes links to the manufacturing and distribution facilities with two nodes in Evreux (France) and connections to the rest of the affiliate offices in Neuilly-sur-Seine (France), Wiesbaden (Germany), Milan (Italy), Barcelona (Spain), and London (UK). Similarly, the U.S. hub in New Jersey connects with the offices in Toronto (Canada) and the headquarters in New York City. The connection between Mount Olive and New York is a T-1 connection that has a capacity of 1.544 Mbps. Figure 8.2 shows a graphical overview of CKCC's international network.

The network's e-mail service is used by approximately 800 account holders—approximately 400 users in Mount Olive and 200 users in New York, 25 users in each of the other countries' offices. As basic data

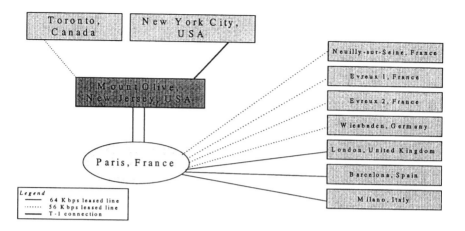

FIGURE 8.2 Graphical overview of CKCC's international network.

transport service, the e-mail application uses TCP/IP file transfer, which is switched by 11 routers in the nodes connected with the international leased lines.

Demarcation of the Case Research

As demarcation of the network in the case, the total CKCC network is chosen. The number of countries the network is located in (seven), is manageable for analysis, and analyzing the complete network makes it easier to obtain performance indicators.

8.2.2 ORGANIZATION OF NETWORK MANAGEMENT

CKCC's network management organization is located in Mount Olive, NJ. Three people are directly involved in the management of the international network. Staff members responsible for information technology management in other country offices do not directly monitor network performance, but signal problems with international lines to the U.S. network management center, which functions 24 hours a day, 7 days a week.

France Télécom has been contracted by CKCC to perform maintenance on pieces of equipment on its international network. France Télécom is also responsible for the equipment used to back up the leased lines in the European affiliate offices. On these sections, France Télécom performs functions of network management and collects a series of performance indicators.

8.2.3 REGULATORY ENVIRONMENT OF THE COUNTRIES

As the demarcation of the case study includes all the countries of CKCC's international network, the countries to be examined are: Canada, France, Germany, Italy, Spain, U.K. and the U.S. All scores were determined in July 1997.

The scores of the regulatory environment are calculated and presented. These include:

- Legal Framework for Competition (see Table 8.4)
- Regulatory Body (see Table 8.5)
- Competition Active (see Table 8.6)

TABLE 8.4
The Legal Framework for Competition of Countries in the CKCC Network

	Legal Framework for Competition					
	Value Added Services		Basic Data Transfer Services		Infrastructure Services		TOTAL Legal Framework for competition
	Code	Score	Code	Score	Code	Score	
Canada	C	3	C	3	C	3	9
France	C	3	C	3	PC	2	8
Germany	C	3	C	3	PC	2	8
Italy	C	3	M	0	PC	2	5
Spain	C	3	PC	2	PC	2	7
UK	C	3	C	3	C	3	9
USA	C	3	C	3	C	3	9

Note: M = Monopoly = 0 point
D = Duopoly = 1 point
PC = Partial Competition = 2 points
C = Competition = 3 points

8.2.4 TELECOMMUNICATIONS SERVICES OFFERING IN THE COUNTRIES

The selected countries were then examined, so the telecommunications services offering matrix could be created as represented in Tables 8.7 and 8.8. The totals on each row, are summed up and shown in Table 8.9 to give a total overview of the telecommunications services offering.

Table 8.9 shows the sums of the respective price and area values.

TABLE 8.5

The Regulatory Body of the Countries in the CKCC Network

					Regulatory Body			
	1. Independence/ Enforcement power	2. Licensing process	3. Equal Access to network Infrastructure	4. Price of interconnection	5. Fair competition regulations		6. Number portability	TOTAL Regulatory Body score
					Price Regulations	Universal Service		
Canada	3	2	2	1	0	1	0	9
France	3	3	3	2	0	0	0	11
Germany	2	2	3	1	0	0	0	8
Italy	2	2	0	2	0	0	0	6
Spain	2	2	2	0	0	0	0	6
UK	3	3	1	1	1	0	3	12
USA	3	3	2	3	1	1	3	16

Note: For scoring legend, see section 3.2.

TABLE 8.6

The Amount of Active Competition in the Countries in the CKCC Network

	Active Competition						
	Value Added Services		Basic Data Transfer Services		Infrastructure Services		TOTAL Competition Active
	Number	Score	Number	Score	Number	Score	
Canada	> 2	3	> 2	3	> 2	3	9
France	> 2	3	> 2	3	> 2	3	9
Germany	> 2	3	> 2	3	> 2	3	9
Italy	> 2	3	0	0	> 2	3	6
Spain	> 2	3	> 2	3	> 2	3	9
UK	> 2	3	> 2	3	> 2	3	9
USA	> 2	3	> 2	3	> 2	3	9

TABLE 8.7
Telecommunications Services Offering Matrix for Countries of the CKCC Network (VPN, ISDN and PSTN Services)

	VPN					ISDN					PSTN				
	A	S	P	S	Total VPN	A	S	P	S	Total ISDN	A	S	P	S	Total PSTN
Canada	+	3	L	3	6	-	1	M	2	4	+	3	M	2	5
France	o	2	M	2	4	+	3	M	2	5	+	3	M	2	5
Germany	o	2	M	2	4	+	3	M	2	5	+	3	M	2	5
Italy	o	2	M	2	4	+	3	M	2	5	+	3	M	2	5
Spain	o	2	M	2	4	o	2	M	2	4	+	3	M	2	5
UK	+	3	L	3	6	+	3	L	3	6	+	3	L	3	6
USA	+	3	M	2	5	-	1	M	2	4	+	3	M	2	5

A = Area; S = Score; P = Price of telecom services

8.2.5 TOTAL SCORES OF THE INPUT PARAMETERS

The total scores that measure the regulatory environment and the telecommunications services offering for the countries of the CKCC network are reproduced in Table 8.10. The total scores for both the regulatory environment (fifth column) and the telecommunications services offering (right-hand column) and individual column scores will be used as input data for the statistical analysis in Chapter 9.

8.2.6 COST-EFFECTIVE MANAGEMENT OF THE CKCC NETWORK

Cost of Ownership

Most of the costs paid by Calvin Klein Cosmetics are recurring charges like external facilities or maintenance expenses. Capital costs such as the purchase of the network equipment are counted in the depreciation cost, although much of the equipment was purchased more than 5 years ago and is already fully depreciated.

In an overview, the costs of ownership for CKCC, classified into the main categories of the COO model, are

TABLE 8.8
Telecommunications Services Offering Matrix for Countries of the CKCC Network (Leased Lines and PSDN Services)

Telecommunications Services Offering Matrix

	Leased Lines 64 Kbps					Leased Lines 2 Mbps					PSDN				
	A	S	P	S	Total	A	S	P	S	Total	A	S	P	S	Total PSDN
Canada	+	3	M	2	5	+	3	M	2	5	+	3	M	2	5
France	+	3	M	2	5	+	3	M	2	5	-	1	M	2	4
Germany	+	3	M	2	5	+	3	M	2	5	+	3	H	1	4
Italy	+	3	M	2	5	+	3	H	1	4	-	1	M	2	4
Spain	+	3	M	2	5	o	2	M	2	4	o	2	H	1	3
UK	+	3	M	2	5	+	3	M	2	5	+	3	L	3	6
USA	+	3	L	3	5	+	3	L	3	6	+	3	L	3	6

Note: A = area of availability

+ = wide area of availability (80%–100% of populated area) 3 points
o = limited areas of availability (20%–80% of populated area) 2 points
- = little area of availability or more than 12 months' wait for service (less than 20% of populated area) 1 point

P= price of telecommunications services
L = Among the least expensive 20% of the countries (3 points)
M = Not among the cheapest or most expensive (2 points)
H = Among the most expensive 20% (1 point)

S = score

- Hardware/software consisting mainly of depreciation on routers in the network
- Personnel (headcount numbers were identified and industry average personnel costs were used to calculate costs)
- External facilities including, in this case,
  - ◆ Rental and connection fees to several service providers for its international leased lines.
  - ◆ Outsourced services, which include the leasing and maintenance fees for the ISDN connections, fees paid to the customer support center of FranceTélécom, and maintenance costs*

TABLE 8.9
Telecommunications Services Offering Matrix Totals

	Total Area	Total Price
Canada	16	13
France	15	12
Germany	17	11
Italy	15	11
Spain	14	11
UK	18	16
USA	16	15

TABLE 8.10
Summary of the Scores of the Input Parameters

	Input Parameters						
	Legal Framework for competition (1)	Regulatory Body (2)	Competition active (3)	Total regulatory environment (1) + (2) + (3)	Area of availability (4)	Price (5)	Total Telecommunications Services Offering (4) + (5)
Canada	9	9	9	27	16	13	29
France	8	11	9	28	15	12	27
Germany	8	8	9	25	17	11	28
Italy	5	6	6	17	15	11	26
Spain	7	6	9	22	14	11	25
UK	9	12	9	30	18	16	34
USA	9	16	9	34	16	15	31

* The costs of the international network of CKCC exclude expenses associated with the leased lines and other network equipment in place on the French facilities of Evreux, and the office in Neuilly-sur-Seine. While the service charges associated with the leased lines are clearly identifiable as expenses linked to a specific country office, the fees that are paid for the maintenance of equipment under FranceTélécom responsibility and for the maintenance of the routers have been equally divided between the nodes based on the type of equipment they contain.

Cost Relative to Size of the Network

Table 8.11 is a summary of the costs incurred by each country office in the left-hand five columns. To calculate the "cost relative to size" of the network, the total capacity of the network in the various countries is divided by the cost of ownership of the network in each of the countries as shown in the right-hand column.

TABLE 8.11
Cost of Ownership and Cost Relative to Size of the Network

| | Costs of Ownership (In thousands of US$ per year) | | | | | |
	Hardware/software costs (1)	Personnel costs* (2)	External facilities costs (3)	Cost of Ownership (4)=(1)+(2)+(3)	Size of network (kbps) (5)	Cost relative to size of the network (In Kbps per thousands of US$) (5)/(4)
Canada	10.91	10	0	20.91	56	2.678
France	48.15	10	122.20	180.35	624	3.459
Germany	16.33	10	23.43	49.76	112	2.250
Italy	16.33	10	50.85	77.18	128	1.658
Spain	16.33	10	32.61	58.94	128	2.171
UK	16.33	10	22.05	48.38	128	2.645
USA	21.82	200	215.59	437.41	3272	7.480

* *Note*: Personnel outside the U.S. is on an allocated basis. Yearly personnel costs are based on industry average in the U.S. Source: information systems departments.

Performance of the Network

As shown in section 5.1, the assessment of performance of the network preferably uses an SLA (described in section 1.7.1) established between the network managers and the users [McGee 1996]. No written SLA was in place between the users of the network and the management of the network of CKCC. However, some performance indicators are regularly collected for network management purposes, and targets have been set by the management team for the network management organization to monitor the level of performance of the network.

Performance indicators* that were tracked are average utilization, peak utilization, network availability, throughput time, MTTR, and number of trouble tickets per month. Peak utilization and number of trouble tickets per month were not used for further analysis, so four performance indicators are used in the case.

Definitions for the performance indicators are established by CKCC as follows:

$$\text{Average utilization} = \frac{\text{Total number of bytes transported per month}}{\text{Available capacity per month}} \times 100\%$$

$$\text{Peak utilization} = \frac{\text{Total number of bytes transported in busy hour}}{\text{Available capacity per hour}} \times 100\%$$

$$\text{Network availability} = 100\% - \left(\frac{\text{downtime hours per month}}{720\text{hrs (month duration)}} \times 100\%\right)$$

$$\text{Throughput time} = \text{Average time it takes for a TCP/IP packet}$$
$$\text{to travel round trip between 2 nodes}$$

$$\text{MTTR} = \frac{\text{Down time hours per month}}{\text{Number of trouble tickets per month}}$$

$$\text{Number of trouble tickets per month} = \frac{\text{Number of trouble tickets opened in a year}}{12}$$

With the exception of average and peak utilization, performance indicators can be directly related to cost-effective management. For bandwidth utilization, the value of the performance indicators depends on the technology used in the network. While certain technologies allow a network to work efficiently when links are almost entirely filled, other technologies need to reserve additional links in case of a "burst"** in traffic.

In the case of the CKCC network, no connection is ever utilized at more than 45–50%, which leaves enough margin to handle peak demands without loss of data. Therefore, we consider that high utilization on the links is an indication of efficient use of the capacity and therefore positive. A positive relationship between bandwidth utilization and cost-effective management

---

* Because the element management of the European nodes is outsourced to France Télécom, a number of indicators are not collected on the connection established to Canada. The statistical analysis is done with the data presently available and excludes the link to Canada from the USA.
** Burst: an unexpected large amount of data that is to be transported.

of the network can be expected. Only average utilization is used for exploring the relationships. Peak utilization is used only to verify that there is no overload in the network that could impact the quality of the data.

Cost-effective management can be considered to increase when availability of the network increases and when throughput time decreases.

MTTR is calculated using the number of trouble tickets per month and provides information on how problems and failures are managed. The interpretation for cost-effective management is that MTTR and throughput time are expected to have an inverse relationship with cost-effective management of the network. The performance indicators used and their scores per country are presented in Table 8.12, where the second line shows the relationship between the indicator and cost-effective management. Network availability as an indicator signifies more cost-effective management when its value is higher and therefore has a positive relationship. MTTR, on the other hand, has an inverse relationship, since a lower value means more-cost-effective management of the network. Later, the calculations will be done and the indicators with inverse relationship will be converted to ones with positive relationship by taking the reciprocal value.

Cost-Effective Management Calculations

In order to make the performance indicators comparable, the indicators with an inverse relationship are calculated with their reciprocal value* so that for all performance indicators a higher value means a better performance. The absence of SLAs made it impossible to weigh the performance indicators or even compare them with existing requirements. The performance indicators are therefore used in their existing fashion to calculate cost-effective management.

The calculation is based on Formula 5.2, found in section 5.4.

The formula shows that cost-effective management can be calculated by taking the performance indicator, or its reciprocal value in the case of inverse related performance indicators, and multiply that by the cost relative to size of the network, which is the same as size of the network divided by cost of ownership. For CKCC, this means the use of the size of the network or total capacity of the links depicted in Table 8.3 divided by cost-of-ownership data of Table 8.11 and multiplied by the performance indicators of Table 8.12. The resulting cost-effective management is compiled in Table 8.13.

For each country, the output parameter is to be related to the input parameters for each country of the cost-effective management model—the

---

* Reciprocal value is "1 divided by the value."

TABLE 8.12
Performance Indicators for the CKCC Network

Relationship	Average utilization (*1) (*2) (%) positive	Network availability (*3) (*4) (%) positive	Throughput time (In milliseconds) (*3) (*4) inverse	MTTR (hours per month per trouble ticket) inverse
Canada	26	n.a.	n.a.	n.a.
France	23.75	99.93	191	2.4767
Germany	26	99.83	200	3.7167
Italy	22	99.99	200	0.030
Spain	19	99.94	230	5.0167
UK	28	99.91	200	2.5
USA	28.5	99.99	125	1.15

Note: All indicators are a yearly average of monthly indicators collected from June 1996 through June 1997.

(*1) Data for Paris are obtained by taking an average of utilization levels on connections to European nodes, excluding the French facilities in Evreux and Neuilly-sur-Seine. Data on the utilization levels between the U.S. and France are not available.

(*2) Data for the U.S. are obtained by taking an average of the utilization levels on links to Canada and New York. This does not include the utilization level on the U.S.–France connections, since the data are not available.

(*3) (*5) Data for France is an average of the indicators collected on all the connections coming into the Paris hub. This includes all leased lines with Wiesbaden, Milan, Barcelona, London, and Mount Olive. It does not include data collected on the national sections of the network such as the connections between Paris and the other French facilities in Evreux and Neuilly-sur-Seine.

(*4) Data for the U.S. concerns only the leased line between Mount Olive and Paris and does not include performance on the connections to Toronto and New York.

n.a. = not available because of outsourcing to an organization that does not report these indicators.

regulatory environment and the telecommunications services offering. The relationships are tested with the help of statistics that can give us an indication of the validity of the propositions, as is shown in Chapter 9.

8.3 CASE 2: UNITED NATIONS DEVELOPMENT PROGRAMME

The United Nations Development Programme (UNDP) is an organization responsible for granting multilateral aid to developing countries. Its main

TABLE 8.13
Cost-Effective Management of the CKCC Network

	Average utilization (% * Kbps per US$)	Network Availability (% * kbps per US$)	Throughput time (1/sec.*kbps per US$)	MTTR (1/Hours * kbps per US$)
Canada	63	n.a.	n.a.	n.a.
France	82.2	345.7	18.11	1.40
Germany	58.5	224.6	11.25	0.61
Italy	36.5	165.8	8.29	165.80
Spain	41.2	217.0	9.44	0.43
UK	74.1	264.3	13.23	1.06
USA	213.2	747.9	59.84	6.50

Note: n.a. = not available

objectives are to promote sustainable human development, eliminate poverty, protect the environment, create jobs, and advance women in society.

The United Nations Development Programme has a network of 143 offices worldwide, and its headquarters are located in New York. The description of the case reflects the situation in May 1997.

8.3.1 OVERVIEW OF THE NETWORK

International Services Offering

The UNDP network offers data communications services such as electronic mail and Internet access (value added services), and TCP/IP file transfer between country offices (basic data transfer service). Those services use different kinds of basic data transfer services such as leased lines, PSDN and very small aperture terminal (VSAT) satellite connections. Table 8.14 illustrates the different services available to the international network locations of the UNDP.

Layout of the UNDP International Network

The international network of the UNDP revolves around the headquarters in New York City. All the connections are between New York and an end point or regional hub. The network of the UNDP is characterized by the existence of only two other regional hubs, from which traffic gets redirected to local country offices. Both of these hubs are operated by the telecommunications

TABLE 8.14
Services of the International Network of the UNDP

Service Value Level	Services	Locations
	Voice Public Switched Telecom Network (PSTN)	Over 100 country offices
	e-Mail and Internet Access	Tunisia, Chile, Colombia
VALUE ADDED SERVICES	International Virtual Private Network (VPN)	Not available
	International Integrated Services Digital Network (ISDN)	Not available
	Packet Switched Data Network (PSDN)	Netherlands, El Salvador and Malaysia
BASIC DATA TRANSFER SERVICES	International leased lines	Switzerland, Malaysia, Bolivia, Denmark, U.S.
	Satellite leased line services (VSAT)	Albania, Belarus, Bulgaria, Latvia, Lithuania, Moldova, Poland
INFRASTRUC- TURE SERVICES		Not available

operator SITA,* which provides data services to the UNDP including the frame relay and the X.25 connections. SITA has two regional hubs that interconnect with the UNDP network. One is located on the east coast of the U.S. in Long Island, and the other in the Netherlands (Burum).

The layout of the network has evolved as such for two reasons—first, the existence of individual communications budgets for each country offices has meant that decisions over the type of connection were taken directly by the local offices based on their needs and resources; second, this particular architecture has enabled the agency to isolate a problem on the network without affecting communications of other countries' offices. A description of the network is shown in Figure 8.3.

The various capacities of the main international links are described in Table 8.15.

---

* SITA (Societé Internationale Telecommunications Areene) is a telecommunications operator created to operate airline reservation systems and their terminals all over the world. SITA merged in 1998 with Equant Communications.

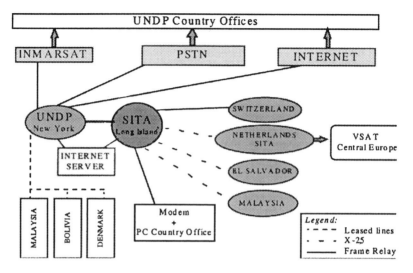

FIGURE 8.3 Graphical overview of the international network of the UNDP.

TABLE 8.15
Capacity of the International Links in the UNDP Network

Link	Capacity
New York–Bolivia (leased line)	64 Kbps
New York–Denmark (leased line)	64 Kbps
New York–El Salvador	9.6 Kbps (X.25)
New York–Malaysia (leased line)	64 Kbps
Internet Connection through T-1 leased line to country offices	1.544 Mbps From 9.6 Kbps to 64 Kbps
New York–Malaysia (backup to the leased line)	9.6Kbps (X.25)
New York–Switzerland	32 Kbps CIR[a] Frame Relay
New York–The Netherlands	64 Kbps (X.25)

a.   CIR (Committed Information Rate) indicates a minimum rate of capacity that is guaranteed on the frame relay service.

Demarcation of the Case Research

Demarcation of the case research was done by selecting a mix of countries. The selection was guided by the need to select countries to which a significant amount of traffic (communication) is routed.

We decided to select the countries with major traffic streams, due either to the importance of the UNDP office in certain countries, for example,

Denmark, Switzerland, and the U.S.; or to the importance of the country as a hub in the network, for example, the Netherlands and the U.S. The selection resulted in a total of seven countries—Bolivia, Denmark, El Salvador, Malaysia, the Netherlands, Switzerland, and the United States.

8.3.2 Organization of network management

The organization of network management is located in New York and is responsible for monitoring the UNDP network. The organization is organized into five units, three of which have an international focus, as is indicated.

- E-mail/Country Office Network Unit (international focus)
- Mainframe Access Control Unit
- Internet Support Unit (international focus)
- LAN Information Unit
- Telephone Unit (international focus)

All units together carry out different activities of network management, which are categorized in OSI functional areas of network management framework. Figure 8.4 presents the organization managing the UNDP network and the number of persons involved in its management. A description of responsibilities is enclosed only for the units that have international focus.

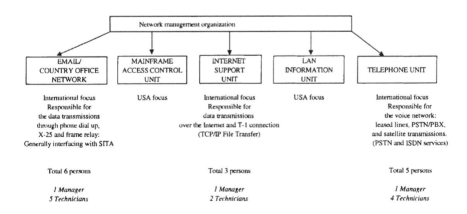

FIGURE 8.4 Network management organization.

In addition to the network management organization in New York, each UNDP country office has, on average, one person who is responsible for managing the network connections with the headquarters in New York. Most of the problems encountered by local offices, such as engineering problems, advisory, or clearance services relating to network operations, are forwarded to the network management organization in New York. SITA provides basic data transfer services as an international operator, and is responsible for managing the performance on the services it delivers to UNDP.

8.3.3 REGULATORY ENVIRONMENT OF THE COUNTRIES

The score of the regulatory environment is determined according to the regulatory environment model described in Chapter 3.

All the information is provided for the period January–June of 1997, which is also the period covered by the performance indicators obtained from the UNDP network management organization. Table 8.16 shows the scores for the legal framework for competition of the countries. Information was available on the Internet about the regulatory bodies for the Netherlands, United States, Switzerland, and Denmark. Information on Malaysia and El Salvador came from local employees in the telecommunications businesses in these countries.

TABLE 8.16
Legal Framework for Competition of the Countries of the UNDP Network

	Legal Framework for Competition						
	Value Added Services		Basic Data Transfer Services		Infrastructure Services		TOTAL Legal Framework for competition
	Code	Score	Code	Score	Code	Score	
Bolivia	M	0	M	0	M	0	0
Denmark	C	3	C	3	C	3	9
El Salvador	C	3	C	3	C	3	9
Malaysia	C	3	C	3	PC	2	8
Netherlands	C	3	C	3	PC	2	8
Switzerland	C	3	M	0	M	0	3
USA	C	3	C	3	C	3	9

Note: M = Monopoly (0 points); D = Duopoly (1 point); PC = Partial Competition (2 points); C = Competition (3 points)

Table 8.17 shows the scores for the countries for the regulatory body.

Table 8.18 shows the scores for active competition in the countries of the UNDP network.

TABLE 8.17
Regulatory Body of the Countries of the UNDP Network

	Regulatory Body							
	1. Independence/ Enforcement power	2. Licensing process	3. Equal Access to network Infrastructure	4. Price of interconnection	5. Fair competition regulations		6. Number portability	TOTAL Regulatory body score
					Price Regulations	Universal Service		
Bolivia	2	2	1	2	0	0	0	7
Denmark	2	3	2	3	1	1	0	12
El Salvador	2	3	2	2	0	0	0	9
Malaysia	2	1	1	1	0	1	0	6
Netherlands	2	3	2	3	1	1	0	12
Switzerland	0	2	0	1	2	0	0	5
USA	3	3	2	3	1	1	3	16

TABLE 8.18
Scores for CompetitionActive in the Countries of the UNDP Network

	Competition Active						
	Value Added Services		Basic Data Services		Infrastructure Services		
	Number	Score	Number	Score	Number	Score	TOTAL
Bolivia	> 2	3	0	0	0	0	3
Denmark	> 2	3	> 2	3	0	0	6
El Salvador	0	0	1	1	0	0	1
Malaysia	> 2	3	> 2	3	0	0	6
Netherlands	> 2	3	> 2	3	> 2	3	9
Switzerland	> 2	3	0	0	0	0	3
USA	> 2	3	> 2	3	> 2	3	9

8.3.4 TELECOMMUNICATIONS SERVICES OFFERING IN THE COUNTRIES

Each country selected from the UNDP network was scored in the telecommunications services offering matrix as shown in Chapter 4. Scores were derived from the Tarifica [1997] reports and OECD [1997] reports. These scores are presented in Table 8.19 for VPN, ISDN and PSTN services and Table 8.20 shows the scores for leased lines and PSDN services. Most of the information was relatively easy to obtain, except in the cases of Bolivia, Malaysia and El Salvador, because of the less frequent study by the consulting firms. Having less frequent reports means that it is more difficult to find reports on all countries that report data on the same period.

TABLE 8.19
Telecommunications Services Offering Matrix for Countries of the UNDP Network (VPN, ISDN and PSTN Services)

	VPN				Total VPN	ISDN				Total ISDN	PSTN				Total PSTN
	A	S	P	S		A	S	P	S		A	S	P	S	
Bolivia	-	1	H	1	2	-	1	H	1	2	o	2	H	1	3
Denmark	+	3	M	2	5	+	3	M	2	5	+	3	M	2	5
El Salvador	-	1	H	1	2	-	1	H	1	2	o	2	H	1	3
Malaysia	-	1	H	1	2	o	2	H	1	3	+	3	H	1	4
Netherlands	o	2	H	1	3	+	3	L	3	6	+	3	L	3	6
Switzerland	o	2	H	1	3	+	3	H	1	4	+	3	M	2	5
USA	+	3	M	2	5		1	M	2	4	+	3	M	2	5

Table 8.21 shows the totals of Area and Price of the telecommunications services offering matrix.

8.3.5 TOTAL SCORES OF THE INPUT PARAMETERS

The total scores of the regulatory environment and telecommunications services offering matrix of the selected countries were compiled. Table 8.22 shows the total scores.

TABLE 8.20
Telecommunications Services Offering Matrix for Countries of the UNDP Network (Leased Lines and PSDN Services)

	Leased lines 64 Kbps				Total VPN	Leased lines 2 Mbps				Total ISDN	PSDN				Total PSTN
	A	S	P	S		A	S	P	S		A	S	P	S	
Bolivia	o	2	H	1	3	o	2	H	1	4	-	1	H	1	1
Denmark	+	3	L	3	6	+	3	M	2	5	+	3	M	2	5
El Salvador	o	2	H	1	3	o	2	H	1	3	o	2	M	2	4
Malaysia	o	2	M	2	4	o	2	M	2	4	o	2	M	2	4
Netherlands	+	3	M	2	5	+	3	M	2	5	o	2	M	2	4
Switzerland	+	3	M	2	5	+	3	M	2	5	+	3	M	2	5
USA	+	3	L	3	6	+	3	L	3	6	+	3	L	3	6

Note: **A = area of availability** + = wide area of availability (80%–100% of populated area) 3 points
o = limited areas of availability (20%–80% of populated area) 2 points
- = little area of availability or more than 12 months' wait for service (less than 20% of populated area) 1 point

P= price of telecommunications services L = Among the least expensive 20% of the countries (3 points)
M = Not among the cheapest or most expensive (2 points)
H = Among the most expensive 20% (1 point)

S = Score

TABLE 8.21
Telecommunications Services Offering Matrix Totals

	Total Area	Total Price
Bolivia	9	6
Denmark	18	13
El Salvador	10	7
Malaysia	12	9
Netherlands	16	13
Switzerland	17	10
USA	16	15

TABLE 8.22
Summary of the Scores of the Input Parameters
of the Countries of the UNDP Network

	Input Parameters						
	Legal Framework for Competition (1)	Regulatory Body (2)	Active Competition (3)	Total Regulatory Environment (1) + (2) + (3)	Area Of Availability (4)	Price (5)	Total Telecommunications Services Offering (4) + (5)
Bolivia	0	7	3	10	9	6	14
Denmark	9	12	6	27	18	13	31
El Salvador	9	9	1	19	10	7	17
Malaysia	8	6	6	20	12	9	21
Netherlands	8	12	9	29	16	13	29
Switzerland	3	5	3	11	17	10	27
USA	9	16	9	34	17	13	30

8.3.6 COST-EFFECTIVE MANAGEMENT OF THE UNDP NETWORK

The calculation of the output parameter cost-effective management of the network should be done similar to the CKCC network, but posed a complication with the tracking of performance indicators in the UNDP network, as will be shown later in this section.

Cost of Ownership

The categories of costs are measured in the UNDP according to the cost of ownership model.

- Hardware/software, consisting of depreciation, rental, and the maintenance of the equipment. The costs are relatively low for the UNDP, as most of the equipment has already been depreciated and maintenance costs are relatively low.
- Personnel, consisting of costs incurred by the New York headquarters. Most network management tasks are done by the headquarters.

- External facilities, which consist of leased lines for data services and voice services.

Cost relative to size of the network

The costs of ownership of the network for each country are now related to the size of the network. Size of the UNDP network is expressed as total capacity of the links. The overview of cost of ownership and cost relative to size, which equals size of the network divided by cost of ownership, is reproduced in Table 8.23.

TABLE 8.23
Cost of Ownership and Cost Relative to Size of the Network

	Costs of Ownership (In thousands of US$ per year)				Size of network (kbps) (5)	Cost relative to size of the network (In Kbps per thousands of US$) (5)/(4)
	Hardware/software costs (1)	Personnel costs (2)	External facilities costs (3)	Cost of ownership (4) = (1)+(2)+(3)		
Bolivia	8	15	47.2	70.2	64	0.9803
Denmark*	13	15	72.3	100.3	64	0.6856
El Salvador	4	15	22.5	41.5	14.4	0.3471
Malaysia	6	15	98.5	119.5	78.4	0.6562
Netherlands	9	15	46	70	64.0	0.9143
Switzerland	7	15	41.8	63.8	32	0.5769
USA	36	820	1143.2	1999.2	1788	0.8947

*The leased line between New York and Copenhagen is a donation of the Danish government to the UNDP. However, in order to avoid introducing a bias in the calculations, the cost of an international leased line between Denmark and the United States was based on the tariffs published by Tele Danmark as of October 1996.

Performance of the network

In the case of the UNDP network, no SLA was available and the organization did not establish internal standards to monitor the activity on the network. Only a limited number of performance indicators were regularly collected. The type of performance indicators available for each UNDP country office were utilization indicators and a record of the incoming and outgoing e-mail traffic.

The indicator average utilization enabled the study of the link capacity and the effective level of utilization of the network connections, which we used as a performance indicator. Average utilization was the only performance indicator available for this study, while the CKCC case had four. Although multiple performance indicators are preferred, this single performance indicator turned out to be adequate. Table 8.24 shows the calculation of the performance of the network, indicated by the performance indicator average utilization.

TABLE 8.24
Performance Indicator and Calculation for the UNDP Network

	Average utilization (1)/(2)	Total number of kbytes transported per minute (1)	Capacity of the network in kbytes per minute (2)
Relationship	positive		
Bolivia	0.066	248	3840
Denmark	0.736	2825	3840
El Salvador	0.806	387	480
Malaysia	0.139	533	3840
Netherlands	0.057	219	3840
Switzerland	0.048	184	3840
USA	0.537	48683	90597

Cost-Effective Management Calculation

Table 8.25 shows the calculation of cost-effective management by multiplying the cost relative to size and the performance indicator average utilization, according to the formula for cost-effective management.

The output parameter cost-effective management has now been calculated, and for each country the output parameter will be related to the input parameters in Chapter 9.

Another performance indicator available was a turnaround time for e-mail traffic. On a monthly basis, the UNDP monitored the amount of time it took country offices to receive a message sent from the New York

TABLE 8.25
Cost-Effective Management

	Cost relative to size of the network (1)	Average utilization (2)	Cost-effective management (1)*(2)
Bolivia	0.9803	0.066	=0.0647
Denmark	0.6856	0.736	=0.5046
El Salvador	0.3471	0.806	=0.2798
Malaysia	0.6562	0.139	=0.0912
Netherlands	0.9143	0.057	=0.0521
Switzerland	0.5769	0.048	=0.0277
USA	0.8947	0.537	=0.4805

headquarters. Percentages of country offices having received the message on the same day, the following day or a number of days after were compiled. However, no numbers were recorded for the individual turnaround time of each country office, which resulted in the decision not to use this data as a performance indicator.

8.4   SUMMARY AND CONCLUSIONS

The cases were carried out in two organizations, the Calvin Klein Cosmetics Company (CKCC) and the United Nations Development Programme (UNDP) according to a standardized approach consisting of six activities. The use of a standardized approach with a predetermined set of activities called for preparation by selecting people in the organization, using questionnaires to get information and using interviews to complete the information and carry out the calculations of the regulatory framework model and telecommunications services matrix to obtain the input parameters. In each of the cases, a network covering seven countries was selected and the input parameters on both sets were determined.

The output parameter, cost-effective management, was calculated after obtaining information on the networks for cost of ownership, size of the network and performance indicator of the network. The performance of the network turned out to be best calculated if an SLA was present between the users of the network and the organization managing the network. In neither of the two cases, however, was an SLA applicable, so performance indicators that were available were used.

In the CKCC case, the four performance indicators were used and as a result four separate calculations of cost-effective management were done. In the UNDP case, one performance indicator was used, resulting in one set of calculations. This availability of performance indicators turned out to be the most essential difference between the two cases and could make the statistical results of the CKCC case study more significant than the results of the UNDP case study.

Now, for each country, the output parameter is to be related to the input parameters in the cost-effective management model: the regulatory environment and the telecommunications services offering and their respective elements. The relationships are tested with the help of statistics that can give us an indication of the validity of the propositions, as is shown in Chapter 9.

9 Results of the Statistical Analysis of the Cases

Step 9 of the research methodology addresses the statistical analysis of the cases, which may show a relationship between the input parameters and output parameter. The propositions of Chapter 7 are tested by doing a number of analyses. Section 9.1 gives an overview of the analysis methods and statistical indicators. Section 9.2 outlines the results for the CKCC case and section 9.3 then shows the results for the UNDP case.

9.1 STATISTICAL ANALYSIS METHODS AND STATISTICAL INDICATORS

To test the relationship between the input parameters and the output parameter, an analysis was used on the data obtained in the case studies. The analysis aims at finding a relationship between an independent variable, in this case the input parameter; and a dependent variable, the output parameter. Each of the countries in the cases forms a set of numbers for input parameters and output parameter. We do not analyze only the total scores of the regulatory environment and the telecommunications services offering as independent variables, but also scores of the individual elements or categories that make up the regulatory environment and telecommunications services offering.

As an additional benefit, the analysis can give insight into the impact of different elements of the regulatory environment and telecommunications services offering on cost-effective management.

A number of regression analyses were done with different independent variables or groups of variables and the explanatory power of each on the output parameter was calculated. A regression with one independent variable is called "simple regression" and a regression with many variables is called "multiple regression." Each case study presents the results of the regressions and makes statements on the relations of the cost-effective management model as shown in chapter 7. Regression analysis can be explained graphically, using a chart with dots that represent the data points, as shown in Figure 9.1.

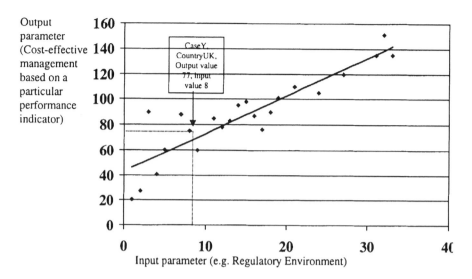

FIGURE 9.1 Regression analysis example.

The points represent combinations expressed in coordinates such as (1,20), (2,28) and (4,41). A line is then drawn through the points to approximate the dots. The line can be straight, in which case it is called a "linear regression line," or have other shapes such as logarithmic, in which case, it will be known as a "logarithmic regression line." To a certain extent, the regression line approximates the datapoints, which can be expressed by statistical indicators that essentially give a measure for the level of approximation.

In this research, each of the dots in the graph represents a country, such as the United Kingdom (UK). The dots are placed at the points where the x-value of the matrix equals the value of the input parameter (here regulatory environment) for that country (for the UK it is 8). The y-value of the location of that dot represents the output parameter—cost-effective manage-ment—with average utilization as indicator (for part of the network in the UK, the value here is 77). Dots are placed for each of the countries in the case.

Out of a common set of statistical indicators we selected two: the r-square and the correlation. Following are descriptions of how each can be interpreted and used.

a. R-Square

R-square, also called coefficient of determination, gives the percentage of total variation of the independent variable that is explained by one or a group of variables, for a simple or a multiple regression respectively. The value of r-square is between 0 and 1. It is also referred to as the explanatory power of the variable or group of variables used in the input.

b. Correlation

The correlation expresses the strength and the nature of the relationship between the independent and the dependent variables. Falling between -1 and +1, it shows a strong relationship when it is a high absolute number. Thus, a correlation of 0 shows no linear association between the input and the output. The negative sign indicates that when the independent variable increases, the dependent variable decreases or vice versa. The positive sign signals that the independent and the dependent variables move in the same direction. In the case of multiple regressions, the multiple correlation explains the overall strength of the relationship between all the independent variables and the dependent variable.

The Tools/Data analysis function of the spreadsheet used for the analysis has a series of statistical analyses, among them regression analysis, which yields several outputs including r-square and correlation. The rest of the output is depicted in an ANOVA (ANalysis Of VAriance) table, which includes several indicators of the variance of the regression analysis. Examples of these indicators are: lower and upper confidence limits, significance value "F" and the so-called "t-stat." Explanation of these indicators can be found in the literature [Knoke 1991].

Statistical Analysis and the Propositions

The propositions are repeated here as a reference for interpretation of the analysis.

Proposition 1—An increase in the score of the regulatory environment in a country results in an increase of cost-effective management of the network in that country.

Proposition 2—An increase in the score of the telecommunications services offering in a country results in an increase of cost-effective management of the network in that country.

9.2   RESULTS OF STATISTICAL ANALYSIS FOR THE CKCC CASE

The relationship between input parameters and the output parameter in the CKCC network was tested by doing regression analyses (regressions) for each of the input parameters, as well as for their individual elements.

In section 8.2.6 the cost-effective management of CKCC was expressed in four values for each country. These were average utilization, availability of the network, throughput time and mean-time-to-repair (MTTR). The performance indicators each have different dimensions and are not additive or combinable into one value. Therefore, four output parameters exist that cannot be combined into one, and so separate regression analyses are done for each of the four. The input parameters are shown summarized in Table 9.1

TABLE 9.1
Summary of the Input Parameters of the CKCC Network

	Input Parameters						
	Legal Framework for competition (1)	Regulatory Body (2)	Active competition (3)	Total regulatory environment (1) + (2) + (3)	Area of availability (4)	Price (5)	Total telecommunicationsservices offering (4) + (5)
Canada	9	9	9	27	16	13	29
France	8	11	9	28	15	12	27
Germany	8	8	9	25	17	11	28
Italy	5	6	6	17	15	11	26
Spain	7	6	9	22	14	11	25
United Kingdom	9	12	9	30	18	16	34
USA	9	16	9	34	16	15	31

Results of Simple Regressions

These results are determined for each of the cost-effective management values that were used to run the regressions. Table 9.2 shows the r-square and correlation values for the regressions of the performance indicators average utilization and availability of the network. A relatively high r-square and a correlation is indicated by a number higher than 0.75.

The regressions for the performance indicators average utilization and availability of the network show a high correlation and r-square value for

TABLE 9.2
R-square and Correlation Values of Simple Regressions of Average Utilization and Availability of the Network in the CKCC Network

Input parameters			Relationship with the output parameter for each of the performance indicators			
			Average utilization r-square	Average utilization correlation	Availability of the network r-square	Availability of the network correlation
Regulatory environment			0.6462	0.8038	0.6327	0.7954
• Legal Framework for competition			0.2979	0.5458	0.3482	0.5901
• Regulatory Body	Total Regulatory Body		0.8229	0.9071	0.7783	0.8822
	1. Independence/ enforcement power		0.3269	0.5717	0.4080	0.6387
	2. Licensing process		0.4055	0.6368	0.4080	0.6387
	3. Access to network infrastructure		0.0386	0.1965	0.0580	0.2408
	4. Price of Interconnection		0.5332	0.7302	0.5156	0.7180
	5. Fair competition regulations		0.7223	0.8498	0.4157	0.6448
		• Price regulations	0.4869	0.6977	0.9219	0.9601
		• Universal service	0.4524	0.6726	0.7508	0.8664
	6. Number portability		0.4869	0.6977	0.4157	0.6448
Competition active			0.1122	0.3350	0.1364	0.3694
Telecommunications Services Offering			0.0722	0.2687	0.0391	0.1982
	Area of availability		0.0326	0.1805	0.0073	0.0856
	Price of telecommunications services		0.0900	0.3000	0.0661	0.2572

the regulatory body, of which the category fair competition regulations shows a significant r-square value and correlation. Other categories of the regulatory body also have a relatively high correlation. The telecommunications services offering does not show high values of r-square and correlation for the performance indicators average utilization and availability of the network. Table 9.3 shows the regression analysis for the performance indicators throughput time and MTTR.

TABLE 9.3
R-Square and Correlation Values of Simple Regressions of Throughput Time and MTTR in the CKCC Network

Input parameters			Relationship with the output parameter for each of the performance indicators			
			Throughput time R-Square	Throughput time Correlation	MTTR R-Square	MTTR Correlation
Regulatory environment			0.9120	0.9550	0.5057	0.7111
• Legal Framework for competition			0.8803	0.9382	0.7345	0.8570
• Regulatory Body	Regulatory Body		0.8207	0.9060	0.2053	0.4531
	1. Regulatory Body independence and enforcement power		0.7184	0.8476	0.1855	0.4307
	2. Licensing process		0.7184	0.8476	0.1855	0.4307
	3. Access to network infrastructure		0.0613	0.2477	0.5937	0.7705
	4. Price of Interconnection		0.1083	0.3291	0.0685	0.2618
	5. Fair competition regulations		0.6123	0.7825	0.0695	0.2637
		• Price regulations	0.7406	0.8606	0.0872	0.2954
		• Universal service	0.2653	0.5151	0.0278	0.1669
	6. Number portability		0.7406	0.8606	0.0872	0.2954
Competition active			0.4240	0.6512	0.9988	0.9994
Telecommunications Offering			0.6848	0.8275	0.1081	0.3288
Area of availability			0.4874	0.6982	0.0763	0.2762
Price of telecommunications services			0.6712	0.8192	0.1065	0.3263

Regression analysis for the performance indicators throughput time and MTTR shows a strong relationship of the regulatory environment with the output parameter. Of the regulatory environment, the element legal framework for competition shows strong correlation and r-square values. The regulatory body, also which shows a strong relationship in Table 9.2, is strong in relationship with throughput time, but not so much in relationship with MTTR. Regressions analyzed for different output parameters provide us with a wide variety of results that require caution when interpreting them for general conclusions. There are, however, some observations that we would like to make for further reference.

- All the input parameters have a positive correlation with the output parameters, which confirms our confidence in the propositions.
- The regulatory environment has a stronger relationship with cost-effective management than the telecommunications services offering does. This is indicated by the higher r-square and correlation values for regulatory environment than for telecommunications services offering in Tables 9.2 and 9.3.
- Out of the elements of the regulatory environment, the regulatory body shows the strongest relationship with cost-effective management; legal framework for competition is secondary, but still relatively strong. They are both represented by a number greater than 0.75 in Tables 9.3 and 9.4.
- Regulatory body categories that show strong reltionships are category 1–regulatory body independence and enforcement power; category 2–speed and openness of the licensing process; and category 5–fair competition regulations.
- Category 4 of the Regulatory Body, access to network infrastructure, shows the weakest relationship, with lower values for the statistical indicators.
- Out of the telecommunications services offering, price of telecommunications services shows a stronger relationship than area of availability.

9.3 RESULTS OF STATISTICAL ANALYSIS FOR THE UNDP CASE

As in the statistical analysis of the CKCC case, the relationship between input parameters and the output parameter in the UNDP network was tested doing statistical analysis for each of the input parameters. The input parameters are summarized again in Table 9.4.

The output parameter, cost-effective management is, for this case study, based on one performance indicator, average utilization. The results of the regressions are presented in Table 9.5

The regressions analyzed and presented in Table 9.5 show that some of the variables have a strong relationship with cost-effective management.

TABLE 9.4
Summary of the Scores of the Input Parameters
of the Countries of the UNDP Network

	Input Parameters						
	Legal Framework for Competition (1)	Regulatory Body (2)	Active Competition (3)	Total Regulatory Environment (1) + (2) + (3)	Area Of Availability (4)	Price (5)	Total Telecommunications Services Offering (4) + (5)
Bolivia	0	7	3	10	9	6	14
Denmark	9	12	6	27	18	13	31
El Salvador	9	9	1	19	10	7	17
Malaysia	8	6	6	20	12	9	21
Netherlands	8	12	9	29	16	13	29
Switzerland	3	5	3	11	17	10	27
USA	9	16	9	34	17	15	31

However, there are no r-square values or correlations of more than 0.75, indicated by the absence of shaded cells in Table 9.5.

The total score for regulatory environment, the scores for legal framework for competition and regulatory body and the categories of the regulatory body access to network infrastructure and price of interconnection have the strongest relationship with the output parameter.

Similar to the statistical analysis of the CKCC case, the analysis in this case shows that the telecommunications services offering has less explanatory power than the regulatory environment. When taken as a total score, the telecommunications services offering produces an r-square of 0.18 and a correlation of 0.43. Studied individually, area of availability and price of telecommunications services respectively have r-square values of 0.18 and 0.21 with correlations of 0.39 and 0.46.

Although, as in the CKCC case, the number of data points used in the case is relatively low, the values for r-square and correlation show that there is significant explanatory power, which supports the presence of a relationship among some of the input parameters and the output parameter.

TABLE 9.5
Results of Regression Analysis of the UNDP Case

Input parameters			Relationship with the output parameter for each of the performance indicators	
			Average utilization r-square	Average utilization correlation
Regulatory environment			0.3407	0.6562
• Legal framework for competition			0.3777	0.6164
• Regulatory body	Regulatory body score		0.5545	0.7446
	1. Regulatory body independence and enforcement power		0.3262	0.5711
	2. Licensing process		0.3541	0.5951
	3. Access to network infrastructure		0.3455	0.6599
	4. Price of interconnection		0.3952	0.6287
	5. Fair competition regulations		0.0607	0.2464
		• Price regulations	0.0004	0.0220
		• Universal service	0.1658	0.4072
	6. Number portability		0.3201	0.5658
• Active competition			0.0665	0.2580
Telecommunications services offering			0.1880	0.4335
• Area of availability			0.1528	0.3909
• Price of telecommunications services			0.2156	0.4643

Results of Multiple Regressions

Multiple regressions, as discussed in section 9.1, perform analysis combinations of input parameters, which may result in many calculations, as many combinations of input parameters are possible. As a trial, one multiple regression calculation was done for the UNDP case study. Table 9.6 shows a multiple regression that was done using a combination of the two input parameters. The combination of the scores for regulatory environment and telecommunications services offering, which is the combination of the two propositions, explained about 0.75 (r-square) of the output, with a correlation of 0.86, which is relatively high.

TABLE 9.6
Results of Multiple Regressions with Two Variables
in the UNDP Case

Number of input parameters	Input parameters	Relationship with the output parameter for each of the performance indicators	
		Average utilization r-square	Average utilization correlation
2	(The propositions 1 and 2) • Regulatory environment • Telecommunications services offering	0.7464	0.8634

Case Conclusions

As in the CKCC case, the results of the statistical analysis are based on seven countries and should be treated with care. Cases may be studied with more data points in order to get a more confident assessment. From the available results, the following observations are possible.

- Regulatory environment shows a relatively strong relationship with cost-effective management, and regulatory body is the element with the strongest relationship.
- All the input parameters have a positive correlation with the output parameter, which confirms that an increasing input parameter relates to an increasing output parameter.

9.4   SUMMARY

Analyzing two cases using statistical analysis has given us insight into possible relationships between input parameters and the output parameter.

While conducting the case studies, we encountered the importance of a strict demarcation of the network to be studied. This includes establishing exactly which hardware and software is part of the network and which are, for example, to be considered peripherals. The calculation of the input parameters posed few difficulties, as long as consistent data sources were used. The sources should also be current to the date of the case study. Some sources, such as consulting companies, do surveys of different countries at different times of the year, which could make the data less comparable. The calculation of the output parameter posed more problems. Cost-effective management was calculated as stated in Chapter 5 using the cost of ownership model, but actually obtaining the performance indicators proved to be challenging. The calculation requires performance indicators, which are prescribed in an SLA. An SLA, however, was not present in any of the cases under study.*

The CKCC case study shows that the company uses four performance indicators, which the users had indicated were essential, but with no predetermined, required performance level. The performance indicators were used in the formulas exactly as they were tracked, but with calculating the reciprocal value of the indicators that have an inverse relationship with performance. Use of the four performance indicators, with no required level, made it necessary to do four individual calculations of cost-effective management, which resulted in four different sets of regression analyses.

In the UNDP case study, some countries proved more difficult from which to obtain information on the regulatory environment and telecommunications services offering. Only one performance indicator was used by the users, the utilization of the network. This resulted in one set of regression analyses.

In both cases, the correlation values are positive and therefore show all positive relationships, which strengthens us in our confidence in the propositions.

R-Square and correlation values of 0.75 or more were considered to show strong relationships and, in particular in the CKCC case study, several of these strong relationships appeared.

The score for the regulatory body and in particular several categories of the regulatory body, show particularly high r-square and correlation. The categories are regulatory body independence and enforcement power, speed and openness of the licensing process and fair competition regulations. We should, however, caution the interpretation of these results, as they are based on a limited sample. Further research, as addressed in step 10 of the research methodology depicted in Chapter 10, may be done in this area to develop a more detailed analysis.

---

* It appears many international organizations that have internal network management organizations have not reached the point where they would formalize the relationship between the departments responsible for network management and the users.

10 Conclusions, Practical Recommendations, Cost-Effective Strategies, Future Steps

In the earlier part of the 10-step research methodology, the research project started with developing the research question and ended with a series of conclusions on application of a cost-effective management model. This chapter depicts step 10 of the research methodology, combines the research results of earlier steps, and derives conclusions for answering the research question and the strategic questions of Chapter 1. Several of the conclusions are then used to develop a checklist for management of international networks. Also, practical recommendations for management of an international network are presented, areas for future research are determined, and the total research project is summarized.

10.1 CONCLUSIONS: RESEARCH RESULTS

The Research Question

In order to analyze the research results, we started from the research question to evaluate how we answered it.

The research question, as defined in section 1.5, asks the following.

What heuristic relationships can be found between input parameters "regulatory environment for telecommunications" and "telecommunications services offering" of a country in which part of the network operates and the output parameter "cost-effective management of the network?"

The research question was approached with the 10-step methodology, which we have come to know in a fairly detailed way. In Chapters 3, 4, and 5 the input parameters and output parameters were developed and depicted so they could be quantified. In Chapter 7, heuristic relationships were proposed and formulated in the following propositions, which were then supported by explanation with heuristics and by a statistical analysis of findings in cases.

Proposition 1—An increase in the score of the regulatory environment in a country results in an increase of cost-effective management of the network in that country.

Proposition 2—An increase in the score of the telecommunications services offering in a country results in an increase of cost-effective management of the network in that country.

The propositions are analyzed with heuristics, as well as with statistical analysis in steps 7 and 9 of the research methodology. Analysis with heuristics included the invention of heuristic intermediate factors, which help explain the relationships in the propositions with existing literature. The analysis with heuristics did confirm the propositions.

Statistical analysis, depicted in Chapter 9, confirmed the propositions as well, although it should be stated that it was based on only two case studies with seven countries each and therefore does not include sufficient data points for a thorough analysis. The statistical analysis is therefore particularly useful as a practical application of a methodology that can be used in future cases to create a more confident assessment of the relationships between the input parameters and the output parameter. The statistical analysis did give more insight into the relationships than just the confirmation of the propositions.

The Strategic Questions

The basis for the research questions were the following strategic questions for an international organization that were developed in section 1.3.

1. How should a country be chosen for establishing a new part of the international network, assuming the country choice is flexible?
2. How should an international network topology with border-crossing links be designed so that the international network can be managed cost effectively?
3. How should a country be chosen as a location for a "hub" to concentrate international traffic between countries in the most cost-effective way?

As steps 8 and 9 confirmed our confidence in the validity of Propositions 1 and 2, the answers to the strategic questions should be:

1. Conduct an analysis of the potential countries where the new part of the network might be established according to the business needs

of the international organization. The analysis should be done with the models described for the input parameters: regulatory environment and telecommunications services offering in all applicable countries. Choose the country with the highest score on each of the input parameters.

2. Using the observations in section 4.2, international links prove to be more than three times as expensive as comparable non-international links. Therefore, the number of international links should be minimized in the topology of an international network. One way to achieve this is to use "hubbing" to concentrate international traffic. Hubbing is the methodology that provides for concentration points with the purpose of minimizing the number of links and in particular the number of international links, as is shown in examples in Figure 10.1.

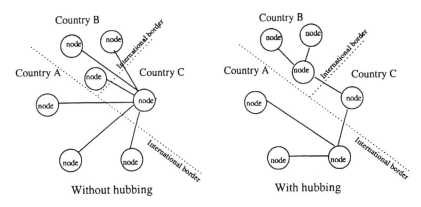

FIGURE 10.1 Examples of an international network with and without hubbing.

The topology without hubbing is shown on the left. It has six links that cross international borders, which results in a high level of costs for external facilities. The topology on the right has only two links crossing international borders and therefore saves cost on international facilities.

3. Hubbing provides for concentration points in a country. When a country is chosen as a hub, many telecommunications services will be purchased in that country because more traffic is coming into and leaving the hub country than other countries. When many services are purchased in a country, by definition the telecommunications services offering in the country is important. One answer to this question is therefore to choose the country with the highest

score for the input parameter telecommunications services offering.

Conclusions from the Statistical Analysis

- All the input parameters have a positive correlation with the output parameters, which confirms our confidence in the propositions.
- The regulatory environment has a stronger relationship with cost-effective management than the telecommunications services offering has with cost-effective management. This is indicated by the higher r-square and correlation values for regulatory environment than for telecommunications services offering in Figures 9.3 and 9.4.
- Out of the elements of the regulatory environment, the regulatory body shows the strongest relationship with cost-effective management; legal framework for competition is secondary, but still relatively strong. They both are depicted with values higher than 0.75 in Figures 9.3 and 9.4.
- Regulatory body categories that show strong relationships are category 1–regulatory body independence and enforcement power; category 2–speed and openness of the licensing process; and category 5–fair competition regulations.
- Category 3 of the regulatory body, access to network infrastructure, shows the weakest relationship, with lower values for the statistical indicators.
- Out of the telecommunications services offering, price of telecommunications services shows a stronger relationship than area of availability.

10.2   RECOMMENDATIONS

Practical Recommendations for International Organizations

Analysis of the cases suggests a series of practical recommendations that can help managers of an international network. They are not based on statistical analysis or quantitative research, but on practical findings during our study:

- Understand international particularities. They do influence management of the international network (see section 6.2) and since they differ in various countries, they should be assessed for every country in the network.
- It is most efficient if a small number of parties manage as many of the service value levels as possible for the international links (see section 2.5).
- High scores for the regulatory environment and for telecommunications services offering are likely to have a positive influence on cost-effective management (see sections 9.2 and 9.3).
- Avoid a high number of international links, as they are more expensive than non-international links and usually involve more parties for their management, which means less control over the quality of the link (see section 4.2).
- Taking these points together in one recommendation for expansion of an international network can be condensed to: Choose countries with high scores for regulatory environment and telecommunications services offering and plan for the right topology of nodes and links, which may reduce the number of international links and/or locate hubs for international links in countries with high scores on the input parameters.

Practical Recommendations for Governments

As depicted in section 1.3, governments are charged with setting regulations for telecommunications. They usually try to provide reliable and low-cost communications for everyone, at the same time making their country as attractive as possible for international organizations to invest in. Making a country attractive for international organizations, according to the propositions, would be helped by influencing the regulatory environment and telecommunications services offering so that the scores of both are as high as possible, although the regulatory environment usuallyis easier to influence by governments than telecommunications services offering. The government should in particular enact regulation that has a strong relationship with cost-effective management, thus having the highest impact in improving the attractiveness of the country so international organizations will buy services in the country, situate hubs, and potentially situate headquarters when the organization is strongly dependent on communications.

The statistical analysis, although it is not completely equivocal, shows that establishing a powerful, high quality regulatory body is particularly important for a high score and the categories independence and enforcement power (category 1), speed and openness of the licensing process (category 2), and fair competition regulations (category 5) should obtain the highest priority for implementation. We therefore formulate the following practical recommendations:

- Open as many services for competition as is possible, given the country can afford a decline in income to the state from telecommunications services.
- Establish a regulatory body with a quick, transparent licensing process and give it enforcement power and independence.
- Let the regulatory body enact regulation that corresponds with the highest scores in the categories.

10.3 COST-EFFECTIVE STRATEGIES FOR MANAGEMENT OF INTERNATIONAL NETWORKS — A CHECKLIST

From the literature and experience with the cases—in particular the explorative case shown in Chapter 6—a checklist can be developed. The checklist is organized by activity in the OSI framework for network management and shows several cost-effective strategies.

Configuration Management

- Check the impact of different legal systems in the countries on the service level agreements with both suppliers and customers.
- Use only services that are considered "available" and "reliable" from the operator in the country. When services are available, but not continuously or don't have the required reliability, then it is better to use a standard, lower-quality service that at least works continuously.
- Standardize domain name addresses for web sites of the company and e-mail addresses in the network in different countries. Either use a general name@company.com e-mail address for all the employees worldwide or use name@company.co (where co is the country) for each of the countries where the network is located.

Fault Management

- Carry management information for international nodes on links separate from the links carrying information.
- Display status information in multiple languages.
- Different time zones of different countries should be taken into account for the establishment of opening hours of the help desks.
- Analyze the support of local equipment providers for the standard equipment in the network and make sure it is sufficient to cover the needs that follow from the SLA.

Performance Management

- Performance indicators should be established that are applicable in multiple countries.
- International links are often expensive and operators use compression to use the bandwidth most efficiently. This, however, may introduce loss of quality and propagation delay. Agree on quality standards with the operators that are providing international circuits.

Security Management

- When introducing multiple operators agree on adequate authentication of users of the network with each of them. This should be covered in the SLA.
- Different laws in different countries can view security breaches (such as intrusion) in different ways. Check the legality of security breaches in all countries in which the network is located and implement extra authentication procedures in countries where security breaches are not against the local law.
- Check if encryption of information is legal in the countries of the network. Encryption should be used on international links wherever legal.
- Check if the encryption algorithms can be exported to the countries of the network

Accounting Management

- Identify costs of the network also on a per country basis, as country tax authorities may require this.
- Do accounting and presentation of the bills in applicable local currencies or in the currency that the customer prefers.

General Strategies

- Minimize the number of parties involved in management of international links at each of the service value levels.
- When more than one party is involved in management of international networks, use a topology that concentrates international traffic and minimizes international links.

10.4   FURTHER RESEARCH

Further research on international networks is highly recommended as a future step, as the subject is of fast growing importance and very little research has as yet appeared in literature (see section 1.7). Some research is found in literature concerning limited areas of the research subject. The regulatory environment, for instance, has been subject to more research recently, but there is none known yet that relates the regulatory environment to management of networks, whereas a prime motivation of governments to change the regulatory environment in a country is to attract international organizations that will invest. As this research indicates, they may be influenced to choose the country in which they invest by the regulatory environment or the telecommunications services offering. A second, separate area where research is being conducted is in developing a more accurate way to determine indicators for cost-effective management. The driver there, however, comes from a completely different background: the recent popularity of providing voice services using data networks, with data services such as basic data transfer services. These data networks use protocols, such as the Internet Protocol (IP). For the transmission of voice, the voice is translated into packets of data to be transported over an IP network, such as for example the Internet (voice over IP). Voice over IP is being adopted by many telecommunications operators and claims to result in more cost-effective management of the networks. As telecommunications operators are about to make big investments in new technology, more research is necessary to guide the type of technology they should implement.

With research being conducted on the regulatory environment and on cost-effective management, the research on input parameter and output parameter is also being further developed, which is a step in the right direction. Apart from developing the input parameters and output parameter, further and more detailed research is necessary to relate the input parameters and output parameter to each other.

While conducting this research, we encountered certain limitations as well as areas where more research could be done, but that fell outside our scope.

- Data used for the statistical analysis in the case studies is based on relatively few observations and therefore the confidence level of the statistical indicators is rather low. An action for further research suggests that more research be done on more cases to provide for more data points and increase the reliability of the results. This will allow the starting values to be refined also.
- The type of telecommunications service used in an international network is not a factor in our examination of the Regulatory Environment, described in section 3.4, as service value levels are used to categorize telecommunications services. However, several regulatory environments have been found that do discriminate between different telecommunications services, even when they are in the same service value level. An action for further research is to refine the model of the regulatory environment to include scores for several different telecommunications services in every service value level.
- The output parameter cost-effective management was measured using a formula developed in section 5.4. This formula is developed because no formula or method was found in literature that was concise and easily applicable. The formula, however, only takes into account three variables, which are each expressed in one numeric value: the size of the network, the cost of ownership of the network and the performance of the network. Using this way to measure cost-effective management, we realize that there are several other factors that could be taken into account that would help make calculations of cost-effective management for different

networks more comparable among each other. An action for further research is to develop a more detailed way to measure the output parameter cost-effective management.

10.5 SUMMARY

International networks fulfill a more and more important role in our society. We use them, for instance, when we get cash from an automated teller machine abroad or when we send someone in another country an e-mail, or simply when we "surf" on the Internet.

An international network is a network that is operational in at least two countries and has a management organization for each of these countries that is responsible for the part of the network that is located in that country.

Managing an international network demands special attention because of the international nature of the network. As the network is in various countries where environments may be different, it should be verified what influence these different environments have on the management. Examples are the different dominant telecommunications services providers in different countries, the fact that there are usually different regulations applicable for offering telecommunications services, and the fact that in different countries a different language may be spoken and understood.

The subject of management of international networks is studied using a 10-step approach.

In the first step, the questions to be answered by the study are formulated.

Literature, trends, and strategic questions heard in international organizations form the basis. Two aspects are chosen and formulated in a research question.

1. The regulation for telecommunications (regulatory environment) in the countries where the network is located
2. The offering of telecommunications services (telecommunications services offering) in the countries where the network is located

The research question researches the relationship between these aspects and the cost-effective management of an international network.

We searched for qualitative relationships between the input parameters and the output parameter by collecting quantitative details of both using statistical models to gain as much insight as possible into possible relationships. Anwering the research question will increase insight into possible effects of these aspects on cost-effective management, so that better decisions can be made for the design and and management of international networks. The heuristic relationships and input parameters found are collectively known as the "cost-effective management model."

Chapter 2, keeping the research question in mind, further developed the theory. An inventory of existing theories on management and international aspects was made and several were judged on their usefulness for modeling input parameters and the output parameter.

Chapter 3 further developed the regulatory environment, which is the subject of many current discussions as countries revising their regulatory environment and international organizations are watching changes closely to see where their next best expansion area might be. The regulatory environment is quantified for a country using three different elements.

1. Legal framework for competition identifies the legal possibilities for competition in different sets of services.
2. Regulatory body is the set of regulations issued by a regulatory body in a particular country.
3. The real state of competition in the country, called "active competition." This last element is a descriptive element, whereas the first two were prescriptive. It actually shows how competition has worked so far and if competitors have been able to obtain any significant market share.

In Chapter 4, the telecommunications services offering in a country was described and quantified. This included an analysis of seven telecommunications services in the country in question. The services were judged for price and availability in particular geographic areas of the country and are depicted in a telecommunications services offering matrix.

After the quantification of the input parameters, Chapter 5 then modeled the output parameter cost-effective management. So far, there are no easy-to-use and adequate measurement methods that were found in theory or practice, so a new model had to be developed. Cost-effective management was defined to include three factors.

1. Cost of management of the network (which can be calculated by an existing method, called the cost-of-ownership method)
2. Size of the network (which can be measured in terms of capacity of all links of the network)
3. Performance of the network (which can be measured with various performance indicators)

Performance indicators are preferably defined in the SLA, which would assure that the users of the network have chosen performance indicators that they consider relevant. In reality, however, many instances have been found where no SLA existed and performance indicators were produced by the organization that managed the network, without regard for the requirements of the users.

A formula was created to calculate a single value for cost-effective management using the three factors.

In Chapter 6, a case study was used to assess the management of an international network in practice and to apply the models for the input parameters in a practical situation. This resulted in a list of attention points that appear as a checklist in Chapter 10.

Chapter 7 described the cost-effective management model, which relates the input parameters to the output parameter. Two propositions were formulated to show the relationships. A more detailed analysis of the case studies was presented using statistical analysis. By means of logical deduction (heuristics) and literature, we concluded that a higher score on each of the input parameters leads to more cost-effective management.

As a second method to test the propositions, Chapter 8 presented case studies carried out on actual international organizations: Calvin Klein Cosmetics Company (CKCC) and United Nations Development Programme (UNDP). In both cases, seven countries in which the organizations had links were chosen for the research. Application of the models for the input parameters and the output parameter was straighforward. We should note, however, that an adequate number of performance indicators should be available on the network of the case studies. The CKCC case study had a higher number of indicators at its disposition, which makes for a higher reliability of the outcome of the CKCC case study than the UNDP case study, which was based on a single set of performance indicators.

Chapter 9 showed the statistical analysis of the cases with the help of linear regression. Linear regression delivers outputs in terms of statistical indicators "correlation" and "r-square." The case studies showed no reason to contradict the propositions, which confirmed our confidence in them.

Therefore, the assumption was strengthened that a higher score of the input parameters relates to a higher cost-effective management of a network.

Specific elements of the input parameter that show a strong relationship with cost-effective management are the element regulatory body in general and the regulatory body independence and enforcement power, speed and openness of the licensing process, and fair competition regulations in particular.

Conclusions

The last chapter, Chapter 10, shows conclusions, recommendations, cost-effective strategies, and ideas for further research. The cost-effective management model is developed to learn more about how aspects of international networks influence the management of the network and therefore how we can improve its cost-effective management by influencing the aspects or choosing the countries with the most favorable aspects. For this book, models of the regulatory environment and the telecommunications services offering in a country were made. The models of these aspects result in a quantification of the aspects in a particular country so that a higher score for the regulatory environment and the telecommunications services offering, while other aspects are kept the same, would be expected to result in more-cost-effective management.

Heuristics and statistical analysis have strengthened the confidence in this expectation. Statistical analysis has also shown that in a country with high scores for certain elements of the regulatory environment and telecommunications offering are particularly indicative for high cost-effective management of a network. The elements are *regulatory body independence and enforcement power, speed, and openness of the licensing process*, and *fair competition regulations.*

Knowledge of these relationships can help to support strategic decisions for management of international networks. Examples of these decisions are (1) the choice of the countries into which the network will expand and (2) the choice of the topology of the network, and (3) the countries where the concentration points (hubs) of the network will best be placed.

Knowledge of the influence of the regulatory environment on cost-effective management is also particularly valuable for governments. With some 69 countries admitting competition in the period 1999–2004, under the WTO agreement and others to follow, the creation of an adequate regulatory environment can be an important way for governments to make their country more attractive for foreign investment in industries such as high technology, which are high users of international networks. Also, a positive regulatory environment will be attractive to a government by creating lower prices of services for the benefit of consumers and thus may increase communication among people in a society.

Further research should possible focus on gathering more-reliable and more-detailed data about the relationships by means of more case studies. The models developed in this book should form a standard for the analysis of those case situations.

Appendix:
Network Manager Survey

A network manager survey was done among 17 network managers of international organizations. The organizations are: Philips, DuPont, ING Netherlands, Shell, Nedlloyd Lines, Netherlands post offices, Hewlett-Packard, IBM, DSM, Monsanto, Akzo Nobel, Dutch Railways, Schiphol Telematics, PTT post, AT&T, Ministry of Health and Environment and the Unilever Meat Group. The survey is reported in the following paragraphs.

SURVEY

In this survey we would like to ask you about the telecommunications services your international organization purchases and which general decision criteria you consider to be important in the buying process of telecommunications services.

A Model of Buying Criteria

To illustrate a purchasing process, we have found a model as depicted in A-1 from theory [Easterby-Smith 1993] that gives an overview of the perceived value and possible purchasing criteria for a product or service. The shaded areas are those we consider applicable for telecommunications services. In your buying process you will perceive a certain kind of value of the service you want to purchase. This perceived value depends on a trade-off between five groups of qualities and the price.

Please rate how important the following variables are for your telecommunications network. You have 10 possibilities to rate: from one (totally unimportant) to 10 (very important).

Example: If you want to rate the importance of a variable as totally unimportant, circle the first number:

totally unimportant $\rightarrow$ very important

1 2 3 4 5 6 7 8 9 10

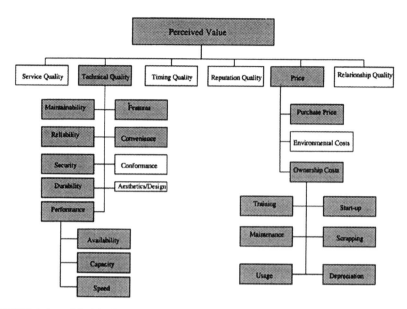

FIGURE A.1   Model of Purchasing Criteria Based on Perceived Value

After finishing the questionnaire, please fax the next two pages and the prepared fax cover page to us.

	totally unimportant/very important
Technical quality (overall):	1 2 3 4 5 6 7 8 9 10
Maintainability:	1 2 3 4 5 6 7 8 9 10
Features:	
Standard features:	1 2 3 4 5 6 7 8 9 10
"Bells and whistles" features:	1 2 3 4 5 6 7 8 9 10
Convenience (overall):	1 2 3 4 5 6 7 8 9 10
Billing options:	1 2 3 4 5 6 7 8 9 10
Outsourcing:	1 2 3 4 5 6 7 8 9 10
Reliability:	1 2 3 4 5 6 7 8 9 10
Security (overall):	1 2 3 4 5 6 7 8 9 10
Authorized access security:	1 2 3 4 5 6 7 8 9 10
Security against private use	1 2 3 4 5 6 7 8 9 10

totally unimportant/very important

Durability:

Technical durability   1 2 3 4 5 6 7 8 9 10

Economical durability   1 2 3 4 5 6 7 8 9 10

Performance (overall):   1 2 3 4 5 6 7 8 9 10

Availability (location):   1 2 3 4 5 6 7 8 9 10

Availability (time):   1 2 3 4 5 6 7 8 9 10

Capacity: 1 2 3 4 5 6 7 8 9 10

Speed: 1 2 3 4 5 6 7 8 9 10

Price (overall):   1 2 3 4 5 6 7 8 9 10

Purchase price/investment: 1 2 3 4 5 6 7 8 9 10

Ownership costs (overall):   1 2 3 4 5 6 7 8 9 10

Training costs: 1 2 3 4 5 6 7 8 9 10

Start-up costs: 1 2 3 4 5 6 7 8 9 10

Maintenance costs: 1 2 3 4 5 6 7 8 9 10

Scrapping costs: 1 2 3 4 5 6 7 8 9 10

Usage costs: 1 2 3 4 5 6 7 8 9 10

Depreciation costs: 1 2 3 4 5 6 7 8 9 10

Other variables:

totally unimportant/very important

_____ :   1 2 3 4 5 6 7 8 9 10

_____ :   1 2 3 4 5 6 7 8 9 10

_____ :   1 2 3 4 5 6 7 8 9 10

_____ :   1 2 3 4 5 6 7 8 9 10

Personal Information

What is your function and what are your main responsibilities?

Function: _____

Responsibilities: _____

Configuration of your company's network

We use the O public network O private network O WVPN (worldwide VPN) network (*2)

Approximately how many end-users are connected? _____

My current operator(s) /PTO(s) is/are:_____

How many sites of your company are connected? In NL:_____
    Worldwide _____

Please fill in the following table:

Kind of traffic		Percentage of total traffic	Yearly expenses
Voice	±	%$	±$
Fax	±	%	±$
Data	±	%	±$
Video	±	%	±$

What are the most urgent problems with your current telecommunications network in order of importance (1=most important, 8=less important)?

1 _____

5 _____

2 _____

6 _____

3 _____

7 _____

4 _____

8 _____

What is the most important function of the international network for your organization?

_____

Are you familiar with national VPN?   Yes/No (*1)

Are you familiar with VPN ?   Yes/No (*1)

(*1) Cross out what does not apply.

(*2) Fill in what does apply.

Glossary of Terms and Abbreviations

Area of availability—Element of telecommunications services offering that indicates in what area of a country telecommunications services are available

Basic Data Transfer Services (BDTS)—Telecommunications services that are between layers 5 and 7 of the OSI model and usually part of a set of services that is not identified as "enhanced services," "value added services," or "infrastructure services" in regulatory environments

Competition active—Element of the regulatory environment that describes how many competitors actually appear when a market is opened for competition

Configuration management—Network management activity for controlling physical, electrical and logical inventories, maintaining vendor files and trouble tickets, supporting provisioning and order processing

Cost of ownership model (COO model)—Model developed by the Gartner Group to measure the cost of ownership of information systems

Cost-effective management—The management of a network in a cost-effective way, i.e., offering a specified set of services/capacities with specified performance requirements for a particular (preferably lowest possible) cost

Cost-effective management indicator—Indicator for the management of a network in a cost-effective way

Cost-effective management model—Model that relates the cost-effective management of a network to the regulatory environment and the telecommunications services offering of the countries in which it operates

Equal access to infrastructure—The right of telecommunications operators to access infrastructure of another (usually dominant) telecommunications operator

175

Fair competition—Competition that does not unreasonably favor a particular party

Fault management—The set of tasks required to monitor and restore faults in order to dynamically maintain agreed network service levels

Infrastructure services—Set of services that operates at layers 1–4 of the OSI model and that usually make available fiber cables, radio spectrum, or wires

Integrated Services Digital Network (ISDN)—a value-added service that is a digital and enhanced version of PSTN. The "basic rate" ISDN service carries two 64-kbps channels and one 16-kbps channel.

Legal framework for competition—This element of the regulatory environment gives an overview of a country's existing laws that regulate the offering of telecommunications services

Licensing process—Process of obtaining a telecommunications license in a country

Management, control, and maintenance (MCM)—The set of activities performed to ensure the continuous fulfillment of the performance characteristics of an information system or network

One-stop shopping—Possibility to obtain multiple services or products from a single supplier

Packet Switched Data Network (PSDN)—A public network that offers the services of transporting data in the form of packets, often using a protocol called X.25, or other possible protocols such as TCP/IP or frame relay

Performance management—Activities required to continuously evaluate the principal performance indicators of network operation, to verify how service levels are maintained, and to identify actual and potential bottlenecks to establish and report on trends for management decision making and planning

Performance of network—Level to which a network meets the performance criteria agreed upon between the users and management organization of a network

Price of telecommunications services—This element of the telecommunications services offering indicates the level of prices offered

Private networks—Networks structured so that no facilities are used that are accessible to the general public

Private line (or leased line)—Dedicated capacity designed to link a single user with requirements in two locations

Public networks—Networks that are accessible to multiple parties of the general public

Public Switched Telephone Network (PSTN)—The publicly available local and long-distance facilities of telecommunications operators

Public Telecommunications Operator (PTO)—A telecommunications operator that is all or in part owned by a government

Regional Bell Operating Company (RBOC)—An operator established after the liberalization of the U.S. market on 1 January 1984 by breaking AT&T into smaller companies

Regulatory body—Element of the regulatory environment that may exist in a country to control the competitive scene of telecommunications operators by issuing and enforcing regulation

Regulatory environment—The set of telecommunications regulations applicable in a country

Regulatory environment model—model that describes and quantifies the regulatory environment in a country

Security management—a set of activities for the ongoing protection of the network and its components, for example, the protection against unauthorized entry of the network, accessing of an application, or transferring of information in a network

Service Level Agreement (SLA)—Agreement between the management organization of a network and its users, which stipulates the required services and performance of the network

Service value level—A classification of a set of services that has a similar "level" in the OSI model

Service value model—A model that describes three different service value levels

Size of the network—A measurement of the scale of a network, for which various methods can be used, such as to sum the total capacity of all links in the part or whole of the network of which the size is to be determined

Subscriber identification and numbering (number portability)—The possibility for a user to keep the same subscriber identification (e.g., telephone number) when switching service providers or when moving within a certain area of a city

Telecommunications services offering (one of the two input parameters for relationships in the research question)—The set of telecommunications services offered in a country, expressed in measurements for price and area of availability of the services

Telecommunications services offering matrix—A matrix that shows the quantification of the telecommunications services offering

Telecommunications services offering score—The score in the matrix that shows the quantification of the telecommunications services offering

Transaction Control Protocol/Internet Protocol (TCP/IP)—A protocol for the transport of packets of information, which, for example, is used in the Internet

Universal service—The offering of telecommunications service at any point in a country at the same commercial conditions (price and installation time)

Value-added services (sometimes called enhanced services)—Set of services (or service value level), available directly to end-users or organizations, that is upward of layer 5 of the OSI model and that is usually performed using the more strictly regulated basic data transfer services

Very Small Aperture Terminal (VSAT)—A satellite terminal that is easily portable and suitable for establishing, for example, leased line service at any location covered by a special satellite

World Trade Organization (WTO)—Organization of countries that was established to promote trade with the fewest possible restrictions between the countries

Telecommunications operator—an operator that provides telecommunications services and is often licensed to do so in a particular area (country or region)

References

Allott, S., Network management, a core skill for future Telcos, *Telecommunications*, August 1997.

Ananthanpillai, R., *Implementing Global Networked Systems Management, Strategies and Solutions*, McGraw-Hill, New York, 1997.

Arnbak, Jens, C., Technology trends and their implications for telecom regulation, *Telecom Reform*, edited by William H. Melody, DTU, 1997 (see ref. Melody 1997).

AT&T, Year report 1997, Corporate communications, Basking Ridge NJ, USA, 1997

Bangemann, Martin, The need for an international charter, *Telecom Interactive '97*, 8 September 1997.

Bartholomew, Martin F., Successful Business Strategies Using Telecommunications Services, Artech House Inc., 1997.

Bishop, T., Operating a seamless global network, *Telecommunications*, international ed., vol. 30, issue 1, January 1996.

Black, U.D, *Data Communications and Distributed Networks*, 3rd ed. Prentice-Hall, 1993.

Cave, M., *New developments in Telecommunications Regulation*, Ch. 8 in [Lamberton1997], Elsevier, 1997.

Cave, M., Alternative telecommunications infrastructures: their competition policy and market structure implications, OECD Competition and Consumer policy division, 1995.

CCTA, Information Technology, Infrastructure, Library, HMSO, London, 1990.

Cole, B.G. After the Breakup, Assessing the New Post-AT&T Divestiture Era, Columbia University Press, New York, 1991.

Clifford Chance, Telecommunications regulations in the Netherlands, internal research, Oct. 1997.

Dalen, E. van, Demand for VPN by multinational corporations, AT&T/Lucent Technologies publication, PO Box 1168, 1200 BD Hilversum, the Netherlands, 1995.

Datacommunicatie atlas, Ministerie van Binnenlandse zaken, The Hague, Netherlands (in Dutch),1990.

Deacon, Graham, and James (law firm), Regulation in Australia, recommendations document, 1997.

Drake, W. E. Noam, The WTO deal on basic telecommunications, Telecom policy, vol. 21, no. 9/10, 1997.

Easterby-Smith, M. et.al., *Management Research, an Introduction*, SAGE Publications Ltd. 1993.

Eekeren, van P., Kosten en baten van computernetwerken, Kluwer Bedrijfswetenschappen (in Dutch), 1996.

Elixmann, D. and H. Hermann, Strategic alliances in the telco services sector, challenges for corporate strategy, paper at the ITS conference, June 16–19, 1996, Seville, Spain.

Eurodata Foundation, Comparison between international leased line tariffs, Public Network Europe, Feb.1996.

Franx, W., Number portability, *Telecommunications Handbook*, Kornel Terplan, Ed., CRC, 1999.

Frieden, R., *International Telecommunications Handbook*, Artech House, 1996.

Gartner Group, A guide for estimating client-server costs, Gartner Group, 1994.

Grover, V., M. Goslar, and A. Segars, Adopters of Telecommunications Initiatives, a profile of progressive U.S. corporations, *Int. J. of information management*, Vol. 15, No. 1, pp 33-46, 1995.

Hammond, A., Universal service in telecommunications, *Telecommunications Handbook*, Kornel Terplan, Ed., CRC, 1999.

Hemmen, van L.J.G.T., Modelling Change management of evolving heterogeneous networks, Ph.D Thesis, Delft University of Technology, The Netherlands, 1997.

Hendriks, J.C.W., M. Looijen, Management van netwerken met een sterk gedistribueerd gebruik, (in Dutch) IT management select, 1997.

Heywood, P. and Ch. Eng, International service providers, the best in the world, *Datacomm*, vol 26, no 6, McGraw-Hill, 1997.

ISO/IEC 7698/4 The ISO/OSI reference model and network management.

Johnson, J. T., International telecom: know how to negotiate, *Data Communications,* June 1997.

Kind, R, M. van Leeuwen, H. Uppelschoten, Managing de kosten van IT beheeer (in Dutch), *Informatie*, May 1996.

Knoke, D., and G. Bohrnstedt, *Basic Social Statistics*, F.E. Peacock, 1991.

Korthals, Altes W., Regulation instruments, in *Telecommunications Handbook*, K. Terplan Ed., CRC, 1999.

Lamberton, D., *The New Research Frontiers of Communications Policy*, Elsevier, 1997.

Liebmann, L., Global connections, *Communications Week*, Sep. 11, 1995.

Loe, D., Business networking: gaining the competitive edge, *Telecommunications*, Dec. 1994.

Looff, L. de, A model for information systems outsourcing decision making, Ph.D. thesis, TU Delft, 1996.

Looijen, M., Beheer van informatiesystemen Kluwer Bedrijfswetenschappen, Deventer, (in Dutch) The Netherlands, 1995.

Looijen, M. *Information Systems, Management Control and Maintenance*, Kluwer, The Netherlands, 1998.

Looijen, M., G. P. van der Vorst, *De rest van de ijsberg*, SURF, WTR, Samsom, (in Dutch) The Netherlands 1998.

McCreary, J., Global Telecommunications services, Strategies for major carriers, *J. of Inf. Syst. Manage.*, Spring 1993.

McGee, K, The consequences of service level agreements, Gartner Group publications, January 24, 1996 (www.gartnerweb.com).

Melody, W.H. (Ed) Telecom reform; principles, policies and regulatory practices, DTU Lyngby, 1997.

Mersel, R.J., Management of distributed data in a distributed environment, Delft University of Technology, 1995.

Michalecki, R., Benchmarks for your telecom operation, *Communications News*, January 1995.

Molony, D., Users mount assault on "high-cost" leased lines, *Communications Week*, 15 February 1999.

Morken, C., How to benchmark your way to better telecom, *Communications News*, July 1994.

Noam, E., *Telecommunications in Europe*, Oxford University Press, 1992.

Noam, E., *Telecommunications in the Pacific Basin: An Evolutionary Approach*, Oxford University Press, 1994.

Noam, E., (Ed.), *Telecommunications in Western Asia and the Middle East*, Oxford University Press, 1997.

Obsitnik, P., Managing the workgroup: common workgroup problems and their solutions, in *3TECH*, vol. 5, nr. 4, oct. 1994.

OECD, Communications outlook, OECD, Paris, 1993, 1995 and 1997.

Oliver, Charles, the Telecommunications Act of 1996, 100 A.B.A. Sec. Science and Technology, *Bulletin of Law/ Science & Technology* 5 (Dec. 1996).

Oliver, Charles, The information superhighway, trolls at the tollgate, *Federal Communications Law Journal*, Dec. 1997.

Oliver, Charles, WTO agreement on basic telecommunications services and fcc implementation, Winter 1998.

Passmore, D., Setting performance expectations, *Business Communications Review*, Dec. 1996

Porter, M., Competitive advantage, *The Free Press*, June 1998.

Research International, Voice of European business, telecom choice, attitudes and expectations, BT and Research International, United Kingdom, 1997.

Roussel, A., The new generation of international carriers, Gartner Group publications, August 16, 1996.

Sandbach, J., International telephone traffic, callback and policy implications, *Telecommunications Policy*, vol. 20 no. 7, 1996.

Schaffers, J., *Options for Competition and Regulation Policy in Telecommunications Markets*, TNO Apeldoorn, the Netherlands, 1994.

Scheele, M. Refile, in *Telegeography*, International Telecommunications Union, Geneva, 1998.

Sisson, P., The new WTO telecom agreement, opportunities and challenges, *Telecommunications*, September 1997.

Tanenbaum, *Computer Networks*, Prentice-Hall, Englewood Cliffs, NJ, 1996.

Tempelman, J., Dutch telecommunications, *Telecommunications Policy*, vol. 21, No. 8, pp. 733-742, 1997.

Terplan, K., *Communication Networks Management*, 2nd ed., Prentice-Hall, Englewood Cliffs, NJ, 1992.

Terplan, K., *Benchmarking Effective Network Management*, McGraw-Hill, New York 1995.

Terplan, K. (Ed.), *Telecommunications Handbook*, CRC, 1999.

Tice, W., and W. Shire, One-stop telecom: the new business model, *Telecommunications*, p. 63–64, April 1997.

Treacy, Cost of network ownership, Index Group 1989.

TMN, Telecommunications Management Network, ITU-T recommendation M.3000.

Tuthill, L., The GATS and new rules for regulators, *Telecommunications Policy*, vol. 21, No. 9/10, 1997.

U.S. Government, Telecom Act of 1996, Published by the Federal Communications Commission, Washington DC, 1996.

Van den Broek, F., and M. Looijen, Management of international networks, *Int. J. Network Mgmt.*, vol. 7 no. 5, 1997.

Van den Broek, F. and M. Looijen, Cost-effective management of international networks, *Int. J. Network Mgmt.*, vol. 7 no. 3, 1997.

Van den Broek, F. Telecommunications services, in *Telecommunications Handbook*, CRC, 1999.

Van der Vlies, M., The transition from monopoly to competition in Australian telecommunications, *Telecommunications Policy*, Vol. 20, no. 5, 1996.

Van Cuilenburg, J. and P. Slaa, Competition and innovation in telecommunications, Telecommunications Policy v19n8 PP:647-663 Nov. 1995.

Verdonck, W., Network management, a strategy for the future, Verdonck, Klooster & Associates, Zoetermeer, The Netherlands, 1992.

Wijs, C. de, Information systems management in complex organizations, Ph.D Thesis, Delft University of Technology, The Netherlands, 1995.

Yankee Group, Regulatory audit, Telecommunications services markets in Asia, 1996.

Yankee Group, Regulatory audit, Telecommunications services markets in Latin America, 1997.

Yin, R., Case Study Research, Design and Methods, Sage Publications, 1984.

Zarley, C. and J. Rosa, Global inroads, Computer Reseller News, Oct. 27, 1997

Web sites with information on telecommunications regulations and international networks:

```
http:/www.wto.org
http:/www.analysys.co.uk
http:/www.att.com
http:/www.fcc.gov
http:/www.infoworld.com
http:/www.opta.nl
http:/www.kpn-telecom.nl
http:/www.bt.com
http:/www.mci.com
http:/www.ins.com
http:/www.level3.com
```

Index